CAD/CAM IN PRACTICE
A Manager's Guide to Understanding
and Using CAD/CAM

CAD/CAM
IN PRACTICE

A Manager's Guide to Understanding
and Using CAD/CAM

A J Medland and Piers Burnett

Kogan
Page

Authors' Note

We have throughout used the personal pronouns 'he', 'him' and 'his' to refer to individuals such as 'the designer' or 'the manager'; this does not mean that we unthinkingly assume that all such people are likely to be male, and certainly not that we believe such a state of affairs to be desirable. It merely reflects the fact that we feel the convolutions of usage that are required to produce a text that is totally non-sexist often represent a cure that is worse than the original disease.

A Note on the Figures

Except where otherwise noted in the captions which accompany them, all the figures in the book were produced on one of the CAD systems currently in use in the Department of Engineering and Management Systems at Brunel University. We would like to thank Computervision who helped by making the other figures available to us, or by creating them to our specifications.

First published in 1986 by Kogan Page Ltd
120 Pentonville Road, London N1 9JN

British Library Cataloguing in Publication Data

Medland, A.J.
  CAD/CAM in practice: a manager's guide to
  understanding and using CAD/CAM
  1. CAD/CAM systems
  I. Title   II. Burnett, Piers
  670.42'7   TS155.6

  ISBN 0-85038-817-1

Printed and bound in Great Britain by
Anchor Brendon Ltd, Tiptree, Essex

Contents

Introduction

Little more than a decade ago computer-aided design and manufacture (CAD/CAM) was a very esoteric field indeed, not one that was of much practical concern to a manager or industrialist unless his business was on the scale of, say, a major automobile manufacturer or in a field of high technology such as aerospace. Like so much else, this situation was revolutionized by the invention of the silicon chip, the arrival of the microprocessor and the dramatic fall in the cost of computer hardware. Today, CAD/CAM has spread down the market, and down the price scale, to the point at which it is both a feasible and an affordable technology for a wide range of small- and medium-sized companies in areas as various as architecture and general engineering, plastic moulding and consumer electronics.

But the explosion – there is no other word for it – in the variety and capabilities of CAD/CAM systems, and their spectacular climb to the top of the hi-tech hit parade, has placed the potential purchaser and user of the new technology in a difficult position. On the one hand he is assured, not least by the manufacturers of CAD/CAM equipment, that a failure to invest in it will leave his company stranded in the industrial Stone Age. But, on the other hand, if he tries to discover exactly what CAD/CAM is all about, and to determine why and how it could be applied to his particular business, he will quickly find himself deluged with a flood of technicalities, swamped by whole alphabets of acronyms and floundering in a morass of competing claims and conflicting advice.

Between the high-level technical works produced by and for those actually engaged in researching and producing CAD/CAM systems (which will tell the average manager rather more than he either wants or needs to know about the technicalities of the subject) and the glossy manufacturers' literature (which may tell him rather less than he would like to know about the relative merits and demerits of the various systems on offer and their relevance to his problems) there is an almost total void. It is that space which this book is designed to fill. Our intention, in short, is to provide a guide, written in plain language, for the prospective user of CAD/CAM who wants to take a careful and considered look at the implications and the options before committing his money, his staff and his business's future prospects.

In the first part of the book we have set out to provide a layman's guide to CAD/CAM – a clear, non-technical summary of its underlying principles and present state of development. This is a prerequisite, since there is, even among practitioners, considerable lack of clarity about the precise meaning of many of the terms employed, the real advantages and disadvantages of different kinds of system and, above all, the way in which the technology is best employed. As some companies which plunged recklessly ahead have already discovered, it is all too easy to be seduced into purchasing a very expensive collection of 'space age' technology, only to find out, too late, that you have bought the wrong kind of system, that you are using it in the wrong way or, in some instances, that you have no genuine use for it at all.

The second half of the book tackles the question of how CAD/CAM is best deployed, how it can fit into a company's existing operations, and how far those operations have to be rethought and reorganized to ensure that the full potential of CAD/CAM is realized. This last point is a vital one. It is fatally easy to assume that CAD/CAM is merely a modern substitute for traditional methods, or a useful adjunct to them, whereas the fact is that, properly used, CAD/CAM requires a radical rethinking of the entire sequence of operations that takes a product from drawing board to market place.

It would, of course, be foolhardy to pretend that any one book can provide a manager with all the information he needs to go out and buy and install a CAD/CAM system to his best advantage. What we have provided, we hope, is a text that will clarify the issues and problems involved so that the reader knows what questions to ask, what difficulties to anticipate and what sort of further information and advice he requires, as well as where it may be obtained.

Chapter 1

CAD – What is it All About?

With the advent of the cheap, mass-produced silicon chip in the early 1970s the rising tide of information technology which had been steadily seeping into industry for the previous 20 years suddenly became a surging flood. One of the consequences of the dramatic growth in the availability and application of computers which followed has been the opening up of a 'generation gap' separating those who ride the fast-moving waves of 'hi-tech' from those whose feet remain planted upon the terra firma of the pre-silicon age.

In the case of CAD/CAM, those who develop, manufacture, sell or apply the technology – the 'insiders', as it were – are by definition on the 'hi-tech' side of this gap; but many of their potential customers are 'outsiders' who find themselves stranded on the other side, baffled and bewildered by unfamiliar ideas and impenetrable jargon. Insiders are full of enthusiasm for the new technology and the visions of the future which it opens up. Among themselves they will talk with relish of the pros and cons of solid modellers, the use of parametrics or the algorithms needed to handle B-spline curves. But they find difficulty in communicating with outsiders, for whom such monologues might as well be conducted in Sanskrit for all the meaning they convey.

Experience suggests that outsiders are seeking answers to questions which, from the insider's point of view, seem almost embarrassingly naive and elementary. First and foremost, they would like to be told 'what CAD/CAM is all about' – and to be told it in plain language. This is not a demand which insiders find easy to fulfil: partly, no doubt, because they have difficulty in imagining what it is like to be ignorant of ideas which they themselves have long taken for granted, and partly, it must be said, because they, like many other 'experts', cherish their monopoly of expertise.

In our opinion, however, the outsiders have a point! Apart from anything else, a grasp of the basic principles involved in CAD/CAM is the indispensable foundation without which no one can be expected to understand the technology or to appreciate the ways in which it might be applied to their problems. We have, therefore, set out to provide in these first two chapters the sort of explanation which outsiders seem to stand

most in need of. Essentially, we shall try to explain in the simplest possible terms and in the plainest possible language what CAD/CAM is, what
it does and how, in principle at least, it does it.

Those readers who feel confident that they have a firm grasp of these
basics may wish to skip ahead to Chapter 3. But we hope that the majority
will bear with us even if some of the points that we make seem obvious or
simplistic. By first clearing the ground and staking out the territory to be
covered we hope to ensure, even at the cost of some impatience on the
reader's part, that the detailed exploration which follows will be both easier and more rewarding.

Concepts and Descriptions

Before anything, be it a jack-knife or a jumbo jet, can be designed, let
alone manufactured, someone somewhere must have conceived of it –
imagined, even if only in very general terms, what it will do, how it will do
it, what it will be made of and what it will look like. This initial step,
forming the *concept* of an object, can only take place inside a mind and,
so far at least, no computer can contribute anything to it.

A concept may be a very simple idea – no more, perhaps, than the realization that an existing component will suffice to meet a new specification
if it is made of a heavier gauge material. It may also be an extraordinarily
elaborate idea such as, let us say, the space shuttle. A concept may be very
general, or extremely precise, dazzlingly original or utterly mundane.
The invention of a new concept may be the principal role of a designer, as
is, for example, the case with an innovative architect or an engineer who
tackles a completely novel problem; but it may also be a step in the design
process which is so trivial and routine that the designer is barely conscious
of taking it.

But whatever the nature of the concept and however great or small the
degree of creativity involved in generating it, it will depend for its existence upon its creator's knowledge and experience of the 'real world'. Any
object, however original in form or function, will be 'like' some other
existing object or combination of objects and will have some purpose
which can be defined in terms of known objects or events. Even if, for
example, one set out to invent a *totally* new kind of mousetrap, one
would still have to know something of mice – what size they are, where
they live, what they eat, perhaps, or whether they are inquisitive.

This kind of knowledge, apparently so simple because it is the stuff of
everyday life, is something which computers have yet to master. The
existing generation of CAD systems, for instance, have such a limited
knowledge of the real world in which people exist that they cannot understand, in the same sense as their users understand, that some of the shapes
they deal with represent material objects while others represent space in

and around those objects – still less do they have any understanding of the purpose of even the simplest of these objects.

To insiders, this point may seem so obvious as to be scarcely worth making. But experience suggests that it is a source of much misunderstanding, and this is not really surprising. Most people have, after all, got used to coexisting with computers which, in other applications, cope with information that seems far more complex and sophisticated than ideas which any infant can master, such as the fact that boxes are hollow and blocks are solid.

A fairly typical example of the sort of misapprehensions that can exist came to light during a seminar at which senior industrial managers were being familiarized with the potential of CAD/CAM techniques. One of the participants, having watched a CAD terminal being demonstrated in operation, then turned to his mentor with a brisk, business-like instruction. 'Right,' he said, 'tell it to design me a telephone!' To the executive concerned, an 'outsider' in our terms, such a request might well have seemed utterly reasonable – after all, his office probably contained a computer which could deliver annual cash-flow forecasts, predict inventory requirements or provide sales statistics more or less ad lib. But, as an 'insider' knows, today's CAD systems are not only unable to deal with a relatively complex concept such as 'a telephone', they cannot even grapple with those such as 'a lever' or 'a crank' which are the small change of any designer's thinking.

But there is, of course, one kind of concept which a CAD system can 'think about' with a speed and skill which far exceeds that of human beings – a geometrical concept. The machine may be unable to share our understanding of the very 'fuzzy' and amorphous rules which we call 'common sense' or 'experience', but it has an enviably complete grasp of the formal and precise rules which make up constructive and projective geometry. It may know nothing of the hardness or softness of materials, of their texture or weight, of the beauty or otherwise of the objects which the images on the display screen represent or of what purposes they are intended to serve, but it knows a very great deal about tangents, cutting planes and skew surfaces. For this is exactly the kind of knowledge – abstract, mathematically rigorous and rigidly codified – which can be packaged in the form of a computer program.

It is also, of course, the kind of knowledge upon which designers rely in order to perform their principal function – the translation of a concept, which may be vague, ill defined and confused, into a design, a *description* which must be precise, formal and unambiguous.

The Design Process

The design process is the sequence of operations which links a concept, the idea of a thing existing only inside someone's head, to a description, the complete design of that thing which specifies it with sufficient accuracy and in sufficient detail for someone to go away and actually make one of them. It can, roughly speaking, be broken down into the following phases:

Drafting. Creating the various geometrical elements – lines, arcs, surfaces etc – which together make up the graphical description of an object.

Checking and evaluation. At regular intervals the designer will have to 'stand back', as it were, to check that the description that is taking shape actually matches the concept with which he started out and that it is satisfactory in other respects – ie that it meets the necessary specifications and falls within the constraints that have been set, that it can be built, that it will do what is required of it, etc.

Correction and amendment. As and when errors are detected, or the designer decides to try a different solution or to modify some part of the description in order to improve it, it will be necessary to delete and redraft some elements – or, in some cases, to start all over again from scratch.

Analysis. Once the description has reached the stage at which, provisionally at least, the designer is satisfied with it, the chances are that it will have to be analysed in more detail. If the object in question is to form part of some larger mechanism or structure, for example, it may be necessary to produce assembly drawings in order to check that all the components will fit together properly and that they will not interfere with one another. It may also be necessary to calculate the object's resistance to physical or thermal stress, and if this process involves techniques of finite element analysis it may well be the longest and most labour intensive part of the entire process.

Manufacturing information. Finally, it will be necessary to make sure that the description contains all the information – dimensions, tolerances, etc – needed at the manufacturing stage. This may well require the production of additional, supplementary descriptions – alternative views of the object, or cutaway or exploded drawings of an assembly, etc.

Although each of these phases represents a stage in the design process, in the sense that there is a progression from a first stage, through intermediate ones, to a final stage, they do not form a fixed and rigid sequence of steps. Instead, as every designer knows, the process is an iterative or cyclical one in which the different phases intermingle and in which progress occurs only in a very general sense. The flow chart shown in Figure 1 represents the enormously complex and often repetitive process of to-ing and fro-ing between different phases which may underlie the broad forward thrust of the designer's work.

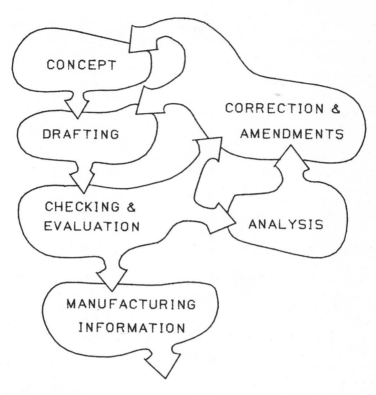

Figure 1. *Although the design process can be broken down into at least six principal phases (and subdivided, if required, into many, many more), it is profoundly misleading to imagine that any piece of design work will follow a linear progression directly from the first phase to the last. On the contrary, as this flow chart shows, the original concept will only become a finished design as a result of a complex series of iterations in the course of which it may well be altered almost out of recognition.*

The Origins of CAD

Computer-aided design, as it is practised today, can be traced back to at least three quite distinct sources, each of which sprang from an attempt to 'computerize' a different part of the design process.

The first of these tributaries, indeed the one which today might be described as the mainstream of CAD, was concerned with automating the creation of graphical information. Those who developed this approach were principally interested in helping the designer during the first three phases – drafting, checking and evaluation, and correction and amendment by devising basic graphics programs that would generate and combine geometrical elements.

Although much of the research that led to today's graphic programs was done in the early 1960s, notably by Dr Patrick Hanratty then working for General Motors Research Laboratories, it did not find much immediate application. At a time when the term computer was still synonymous with one of the vastly expensive giants which we today call a mainframe, designed to perform dozens of functions and service hundreds of users simultaneously, CAD was not a very economical proposition because of the high demands it makes on a machine's processing time and memory capacity. It was not until the arrival of the minicomputer and, subsequently, the microprocessor made it possible to dedicate a single computer to, first, a multi-user CAD system and, later, a single terminal, that graphics programming was able to realize its potential as a general-purpose design tool.

Rather different economic imperatives influenced the development of the second source of today's CAD programming – that concerned with the analysis of designs. The point in this case was that in those industries (like aerospace, say, or semiconductor manufacture) which were extending the frontiers of high technology, the analytical and production techniques which had been developed during the 1950s and 1960s were already stretching human resources to the limit. Only the speed and number-crunching capacity of the computer could make feasible the routine use of, say, finite element analysis. Moreover, since cost considerations were likely to be secondary in these hi-tech areas, the economic constraints on the development of this branch of CAD tended to be relatively unimportant.

In the early days, of course, many of the analytical programs tended to be developed 'in house' to meet the needs of a particular company, and to some extent this is still the case. But a finite element package is now a standard part of most CAD manufacturers' software, and other, more specialized, programming is also beginning to be generally available. The programs designed to simulate the flow of material in a mould (see page 151) or those intended specifically for the design and analysis of pipework are excellent examples of this trend.

The third and final source of CAD development arose out of the effort to use computers to improve and speed up the flow of information from design office to factory floor by building upon the existing numerical control (NC) technology which had been widely adopted by the mid 1960s. This is, of course, the bridge which links CAD to CAM, and we shall be returning to the subject in the next chapter. For the moment, the important point to note is that although many current CAD systems still tend to show clear evidence of having been derived from one or other of these three sources, they all share a common reliance upon the computer's ability to construct and store representations of geometrical *models*.

The model which a graphics program constructs and displays is the CAD equivalent of the traditional, plans-on-paper description which a designer produces with the aid of pencil, paper and drawing board. It is the geometrical model of an object, contained in mathematical form in computer memory, which an analytical program operates upon. And the programs which generate NC data, specifying the path to be followed by a tool, are designed to reproduce in real life the geometrical forms represented by the computer's model.

The nature of a geometrical model is fundamental to virtually all aspects of CAD/CAM and, more than any other factor, is responsible for its advantages over traditional methods. The way in which models are constructed and the uses to which they can be put are, therefore, subjects which we must look at in some detail.

Automated Drafting: Creating a Model

The designer using a CAD system, just like his counterpart sitting at a drawing board, creates graphic descriptions by combining various geometrical elements of which the most basic are lines, arcs and circles.

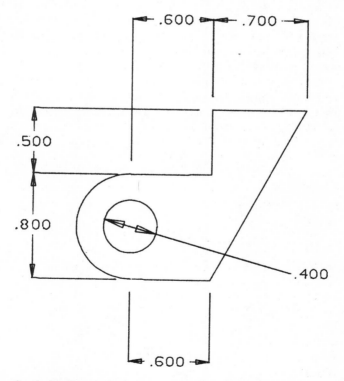

Figure 2. *A simple two-dimensional model made up of seven entitities – five lines, an arc of 180° and a circle. In order to construct this model on a CAD system the designer need only specify the position of each entity and its size – see text for a more detailed account.*

'Drawing' on the display screen, using one of the input/menu arrangements described in Chapter 3, is not very different in principle from drawing on a bit of paper. Consider, for example, the simple, two-dimensional (2-D) model shown in Figure 2. This is made up of seven *entities*: five lines, an arc of 180° and a circle.

In order to construct the model, the designer would start by designating a point, say the centre of the circle, in terms of a set of coordinates along the system's x and y axes. (For convenience, the x axis can be thought of as a horizontal line across the display screen and the y axis as a vertical line from top to bottom of it.) Given the radii of the circle and the arc which is concentric with it, the CAD system can then be instructed to 'draw' these two elements. Further simple instructions will then cause it to generate the five straight lines which complete the model.

Looking at the image of this model as it appears on a display screen or is drawn by a plotter, it is not immediately obvious that it differs in any significant way from the comparable image that might be drawn on a sheet of paper. But if we go on to look at what the system can do with even an elementary design of this kind, it becomes clear that a model, made up of entitites recorded in computer memory has many advantages, from a designer's point of view, over a plan made up of lines on paper.

To start with, once the model has been created it can itself be treated as a single entity. This allows the designer to duplicate and reduplicate it in a variety of ways. Figure 3, for example, was produced by 'mirroring' the original model, while Figure 4 shows how, once the model has been designated as a 'macro', the system can quickly and easily reproduce it in a variety of positions, orientations and combinations.

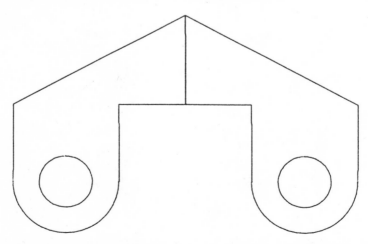

Figure 3. *This image was created by simply instructing the CAD system to 'mirror' the original shape shown in Figure 2, a far simpler process for the designer than laboriously repeating the entire sequence of drafting operations.*

Figure 4. *In order to produce this image the CAD system was first told that the shape shown in Figure 2 was to be treated as a 'macro' and it was then instructed to reproduce that macro in a number of different positions and orientations in order to produce a larger and more complex model.*

Figures 5 and 6 show how new images can be created by 'zooming in' on an area of the model or by using the technique of parametrics (see pages 108-111) to alter the sizes of the component entities while retaining the basic relationships between them.

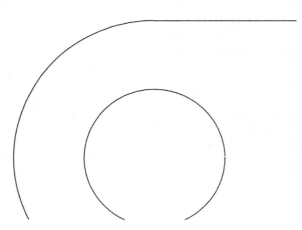

Figure 5. *By zooming in on one small area of an image the designer can inspect it in greater detail or, perhaps, elaborate it by adding details which it would have been impossible to 'draw' clearly while working on the original scale.*

The 'forest' shown in Figure 7 offers a striking example of the way in which parametrics can be applied to a model consisting of just six enti-ties – five lines and a diamond – (itself created from a further four lines) – used over and over again in slightly different forms.

Figure 6. *Each of these images is a variation on the original one shown in Figure 2. Using a CAD system with parametric modelling ability, it is possible to create such variations* ad infinitum *by simply varying the dimensions of the individual entities. In the case of the image in the centre, for example, the radius of the circle was increased so that it was larger than that of the arc while the length of the line joining the bottom of the arc to the line running diagonally across the display space was given a negative value.*

Figure 7. *This 'forest' of trees was created by multiple use of a single macro made up of six entities with constant adjustment of the parameters. The use of such techniques can achieve dramatic economies of time and effort since each repetition of the macro requires the designer to do no more than provide the system with a few elementary instructions.*

These few examples should make it clear that a CAD system can make the initial drafting phase of the design process a great deal easier simply by taking over much of the painstaking work that previously had to be done manually using the traditional paraphernalia of ruler and set square, compasses and dividers. The advantages become even more obvious when correction or redrafting is necessary, for the tedious and messy business of erasing and redrawing lines is replaced by the comparatively painless one of issuing instructions to the system which is capable of drawing, redrawing and re-redrawing without complaint or so much as a trace of a smudge!

Representations and Simulations

But the most significant advantage of a model, as opposed to a plan or set of plans, can be seen when we add a third dimension. For it then becomes clear that, unlike a 2-D plan, which can never be more than a representation of just one aspect (out of the infinity of possible aspects) of an object, the model stored in a CAD system is a *simulation* of that object.

The point may become clearer if we consider a more familiar example of the computer's ability to simulate reality – say a game of video table tennis. Most people will have played such a game at one time or another and will be familiar with the way in which the dot of light, representing the 'ball', 'bounces' backwards and forwards across the screen simulating the behaviour of a real ball. The machine is able to present this illusion of reality because its program contains a set of rules for calculating the motion of the ball (the trajectory along which it should rebound, and with what speed, etc) which are very similar to the rules governing the behaviour of real-life table tennis balls.

In rather the same way, a CAD graphics program contains a set of rules (essentially the rules of projective geometry) which allows it, once it has been given a geometrical definition of an object, to simulate the appearance of that object in a variety of attitudes and according to a wide range of graphical conventions.

To illustrate the point, let us return to the simple 2-D model which we started with in Figure 2 and extend it into a third dimension. We could, for example, project a duplicate of the model a set distance along the z axis (for convenience, again, the z axis can be thought of as a line running back 'into' the display screen from its surface). The result will be two identical models arranged as in Figure 8. (In fact, of course, the models are not stored inside the display unit in graphical form, but in computer memory in mathematical form.)

If we now join up the two models, we will have constructed a single model of a 3-D object, as shown in Figures 9 and 10.

Assuming that the system we are using has 3-D capability, it will now

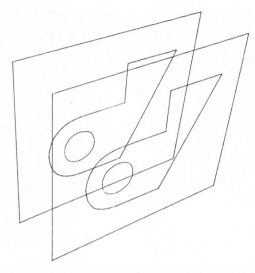

Figure 8. *Adding a third dimension: the original 2-D model shown in Figure 2 can be duplicated in a third dimension by projecting it 'back' along the z axis – pushing it further 'behind' the surface of the display screen, as it were.*

be able to generate and present on the display screen any number of different representations of this single model such as, for example, a set of three-view drawings (Figure 11), a selection of perspective views (Figure 12) or a close-up of one portion of the model (Figure 13).

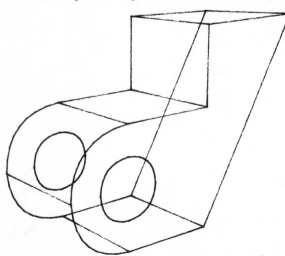

Figure 9. *By 'drawing' six lines which link corners in the original model to the corresponding corners in the new one, a 'wire-frame' model representing a 3-D object is produced.*

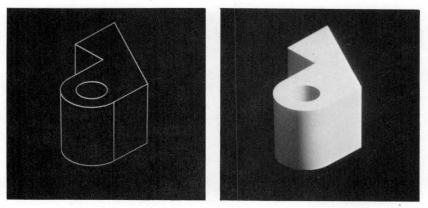

Figure 10. *Removing the hidden lines from the wire frame and adding the horizon line which shows the edge of the curved surface as it would appear from this angle results in a more intelligible image (left) which, in turn, can form the basis of an extremely realistic model (right). (Figure reproduced by courtesy of Computervision.)*

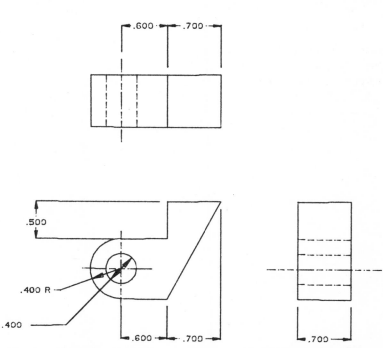

Figure 11. *Once the 3-D model shown in Figure 9 has been created the designer can instruct the system to produce a whole range of different representations of it. This figure, for example, shows a formal, three-view set of engineering drawings complete with dimensions, all of which can be added automatically from the data that was used to construct the original model.*

Figure 12. *A range of perspective views. In order for the system to create these variations on Figure 9, it was only necessary for the designer to specify the desired viewpoint.*

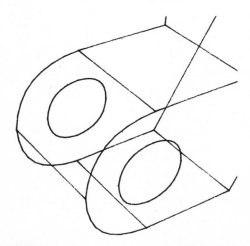

Figure 13. *The designer can 'pan' on to one section of a 3-D model and enlarge it by use of the 'zoom' facility in just the same way as with a 2-D one.*

The fact that views like these, and a virtually infinite variety of others, can be constructed by the machine once it 'knows' the geometry of the model it is dealing with is of enormous benefit to the designer, especially during the checking and evaluation phase. The fact that the designer and the computer continuously interact makes it possible to inspect the model as it develops from any angle, or at any scale or level of detail which is appropriate. In the case of very advanced systems it may even be possible for the designer to 'fly around' the model as if it was being rotated in his hand (or as if he was, literally, flying round it in a helicopter).

When dealing with objects that have complex shapes, or with assemblies containing several separate objects, designers have become accustomed to supplementing the traditional orthogonal views – plan, front and side elevation in the architectural convention – with explanatory and detail drawings. Figures 14, 15 and 16 show some examples of representations which are confusing or ambiguous, together with the clarifications – pictorial, cutaway and exploded drawings – which remove the uncertainty.

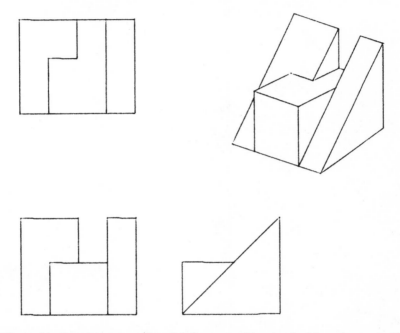

Figure 14. *Supplementary pictorial views are often required in order to clarify points that would otherwise be confusing or ambiguous.*

Using traditional methods, such clarifications are, of course, indispensable even to designers themselves. For once an object attains a certain level of complexity the designer can no more hold a complete picture of it in his head than a writer can remember every detail of the dialogue he gave his characters a couple of hundred pages earlier in the narrative. The designer, like everyone else, must therefore look at the various representations of an object and assemble them, in the mind's eye as it were, into a model of the object. One of the great advantages of a CAD system as an aid for the designer is that it can effortlessly retain, in the eye of its electronic 'mind', a model which is complete and precise and can produce a

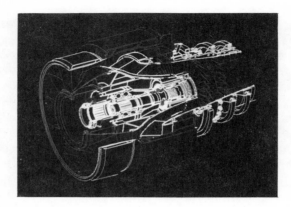

Figure 15. *If an object has a complex internal shape then cutaway drawings will almost certainly be needed to reveal it.*

Figure 16. *As everyone who has had cause to make use of, say, a car maintenance manual will know, exploded drawings can be invaluable for clarifying the relationships between components in even relatively simple assemblies.*

whole range of different representations of that model more or less on demand.

In effect, therefore, once a geometrical model of an object has been created in computer memory, the use of a CAD system allows a designer to 'look at' that object in rather the same way as one might look at an object held in the hand. It can be turned this way and that and the back, the underside, even the 'inside', can be examined; a detail can be inspected in close-up or one can stand back and see the object as a whole; it can be fitted together with other objects, and so on. This makes it possible for a designer to check and evaluate his work far more easily and far more thoroughly than was previously the case. Instead of having to look at a series of 2-D drawings and then try to visualize what something might look like 'in the round', the designer can examine, fiddle about with, modify and experiment with what is, in effect, a simulation of reality.

It must, of course, be remembered that what the system is simulating is only a partial representation of reality. The images which appear on the screen may make it clear what an object might look like in real life and perhaps how it would work, but they will tell the observer nothing about its other characteristics – how heavy it would be, what its surface texture would look like, whether it would float or sink, etc. Basic graphics programming, in other words, is designed to deal only with the geometrical aspect of things.

Analytical Programs: Simulating Performance

In order to analyse and test those characteristics of a design which are not solely determined by its own geometry – its resistance to stress, for example, or its thermal properties – it is necessary to simulate other aspects of reality. The abilities of CAD in this respect are widening very rapidly. In the semiconductor industry, for instance, programs have been developed which, once they have been given the geometrical description of a microchip (its wiring diagram, so to speak), can then simulate not only its electrical characteristics but also its logical ones. We shall come back to these, and other examples, in Chapter 6. For the moment we will concentrate on the two kinds of analysis which are most generally applicable: stress calculations and the simulated operation of mechanisms or their interaction with other mechanisms.

Figure 17 shows the simple model which we used earlier on in the chapter divided up into a large number of separate elements by a computer-generated 'mesh'. Using the technique of finite element analysis it is possible to calculate how any given physical or thermal stress will affect each of the elements and, thus, to establish which bits of a structure would give way or distort under what conditions. Clearly, actually generating a mesh and then performing a separate set of stress calculations for each element

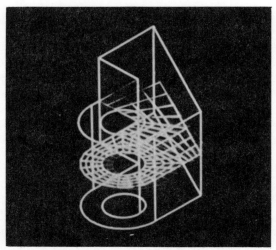

Figure 17. *By automatically generating a mesh which divides a model into a large number of separate elements a CAD system can enormously speed-up the process of analysis. Here the centreline of the object shown in the figure has been 'meshed'. Finite element techniques require that the behaviour of each element under given conditions of physical or thermal stress should be separately calculated, a task which the computer, with its enormous capacity for high speed calculation, is ideally suited to perform. (Figure reproduced by courtesy of Computervision.)*

is an enormously laborious process, but one ideally suited for capitalizing on the computer's sheer speed and number-crunching capacity. It is no exaggeration to say that there are today many design tasks which would be totally impractical were it not for the existence of CAD programming which automatically generates meshes, does the stress calculations and, in some cases, even simulates, on screen, the distortion of an object under stress, as shown in Figure 18.

Traditionally, when dealing with the design of complex mechanisms, engineers have at some stage found it necessary to build one or more prototypes in order to check that something which 'looked right' on paper would actually work in practice. For, however meticulously the theoretical calculations may have been carried out, there will always be some problems that are most easily solved by trial and error – a clearance which, it turns out, should be more generous, a spring that needs to be strengthened, two moving parts that interfere with each other in a fashion that was not foreseen.

CAD techniques, especially those involving kinematics (simulated movement), now make it possible for designers to make their trials and discover their errors without actually having to build a real prototype. Instead, by simulating the operations of mechanisms and their interactions with other mechanisms snags can be identified and remedies tested without ever leaving the CAD terminal.

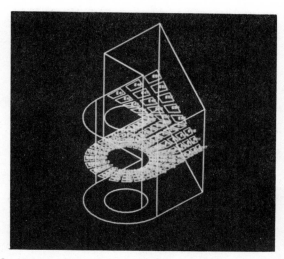

Figure 18. *Many analytical programs based on the use of finite element techniques can also provide graphical representations of the behaviour of a model under specified loads. In this case the finer lines represent the distortions that would occur in the mesh shown in the previous figure. (Figure reproduced by courtesy of Computervision.)*

Figure 19 shows an example of the application of kinematics to the planning of automated production lines. Where a new line involves the use of robots, to weld car bodies for instance, they must be positioned with great care. Obviously, they must not clash with each other or with the moving line and its associated equipment; they also must be able to reach into a number of different positions, some of them probably awkward in the extreme, and they must be able to perform all their allotted tasks within the time one car body remains within their 'reach'. By doing all the planning on a CAD system, and perhaps programming the robots themselves 'off-line' before the plant has even been constructed, much time, trouble and money may be saved.

Summary: CAD Defined

At this point, before we move on to consider the links between CAD and CAM and the way in which CAD can be used to improve the final phase of the design process – the provision of manufacturing information – it is worth trying to summarize what has been said so far by providing a brief definition of 'what CAD is all about'.

Computer-aided design, we may say, is a technique whereby the geometrical descriptions of objects can be created and stored, in the form of mathematical models, in computer memory. A CAD system is a combination of computer hardware and software which facilitates the con-

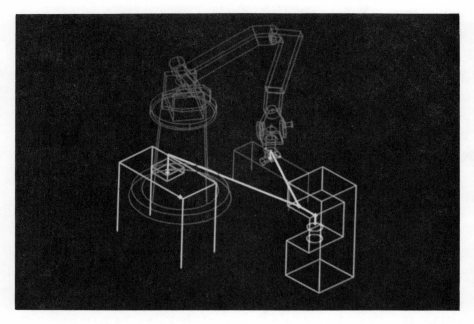

Figure 19. *Kinematic programming makes it possible actually to simulate, on screen, the movement of mechanisms and their relationship to each other. It is now commonplace in the motor industry for robots to be positioned and, sometimes even programmed off-line by the use of such techniques. (Figure reproduced by courtesy of Computervision.)*

struction of such models, and in many cases their analysis too, and which allows a wide variety of visual representations of those models to be displayed.

Chapter 2

CAM – An Introduction

Design and Manufacture: Two Processes or One?

The most elusive part of the term CAD/CAM is that deceptively simple oblique stroke which links (or does it separate?) the two halves. We have already shown, we hope, that CAD has a clear and comprehensible identity, and we hope to show in a moment that the basic principles of CAM are equally easy to grasp – but what of the relationship between the two?

Most of the problems and confusions which arise here have little or nothing to do with the use of computers; they are essentially a consequence of misunderstanding the relationship between the work of the designer (with or without the aid of a computer) and the production process.

Common sense would suggest that design and manufacture are best thought of as totally distinct and separate operations: they are, after all, performed by different people, in different places, at different times, using different tools and different skills. Once a design is complete, surely, the designer's work is done. He simply hands the finished description of the object over to someone else, a representative of 'manufacturing', who takes it away and, using the information it contains as a guide, actually makes the object. Seen in this way, the two processes are quite separate stages in the overall process which transforms a concept into a product, as shown in the flow chart in Figure 20.

This is certainly an attractively orderly way of looking at things. Unfortunately, however, it is not a very accurate or useful representation of what actually takes place when something is designed and manufactured.

Consider, to start with, the point that many designs consist not merely of descriptions of what is to be made, but also of instructions as to how it is to be made. Even if the designer does not tell the manufacturer how he should set about his work, or what equipment he should use, it is still likely that the choice of tools and techniques at the manufacturing stage will have been a major factor in the designer's thinking.

It would be quite unrealistic, for example, if something like a toothbrush, intended for mass production and mass marketing, were to be designed in such a way that it could only be manufactured by machining it

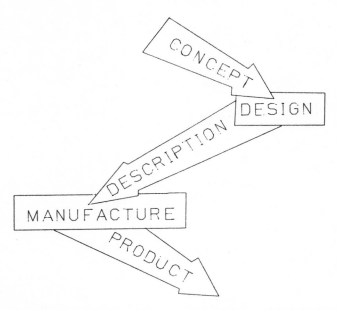

Figure 20. *To the layman common sense might suggest that design and manufacture are two entirely distinct processes, taking place one after the other, as shown in the flow chart. In practice, however, it is much more realistic to look upon them as two aspects of a single process.*

out of solid material. But it would be equally absurd to design a custom-built, one-off product in such a way that it involved reliance upon mass-production techniques such as an assembly line.

In practice, clearly enough, the designer's knowledge of the materials, the equipment and the skills which are available will be a major influence upon the designs that he produces. In fact, the job of producing a formal design of any complexity will almost certainly involve the simulation, if only in the designer's imagination, of some or all of the manufacturing processes. Indeed it is often a large part of a designer's responsibility to plan how something can best be manufactured and to foresee any problems that might arise. Certainly, the case where the designer is confined simply to deciding what should be made while the question of how it should be made is left entirely to the manufacturer is the rare exception and not the general rule.

There are, though, many circumstances in which the situation is reversed: where decisions which are, in theory, matters of design are left to the discretion of those responsible for the manufacturing process. No craftsman in the construction industry, for example, would expect an architect to specify every last little detail of a building's design. On the contrary, a plumber or an electrician will normally expect to do much of the detailed design of an installation as the work progresses. Even in an engineering shop, it may be left to a machine operator to decide when a curve 'looks right' or to a fitter, working with a hand file, to decide, within a set margin of tolerance, when he is satisfied with the way two components go together.

A more accurate way of thinking about design and manufacture, and the way in which they relate to each other, is, therefore, to view them as two aspects of a single continuum. This arrangement can be represented diagrammatically, as in Figure 21.

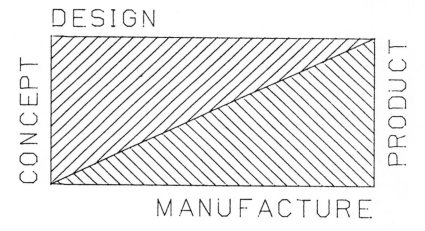

Figure 21. *The design-manufacture continuum. The diagram accurately reflects the fact that* most *design work precedes* most *manufacturing work, but also shows that* some *manufacturing considerations will have to be taken into account right from the start of the design process while, conversely, some design decisions may remain open until the final stages of manufacture.*

There is a constant flow of information in both directions. From the design sector, descriptions and instructions are passed to the manufacture sector; in return, information about the availability of tools and materials, the relative economics of different processes and so on flows from manufacture to design. The line between the two must, therefore, be seen as a permeable and often a 'fuzzy' one. In fact, if we consider the diagram as a way of representing, in very general terms, a relationship that may vary enormously from one example to the next, we must recognize that the dividing line which separates the two processes is not only loose, it is also extremely elastic. Either of the two processes, in other words, may expand until it occupies virtually all the space in the continuum. A couple of contrasting examples will illustrate the point.

Consider, first, the case of a carpenter who has been employed to build a closet in someone's spare bedroom. It may be that his instructions amounted to no more than the vague request that the closet should 'look like' one in another room. In those circumstances the carpenter will probably 'design' the closet more or less as he goes along, taking account of the space which it must occupy and the materials available to him. Diagrammatically, the entire process of designing and manufacturing the closet could be represented by Figure 22, where the manufacture function has virtually squeezed the design one out of existence.

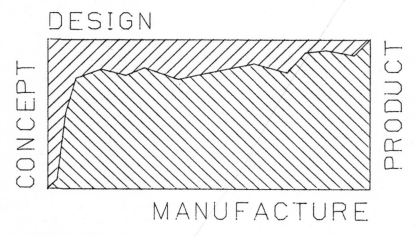

Figure 22. *A craftsman such as a carpenter may virtually design a product 'as he goes along', in which case almost the entire continuum is occupied by the 'manufacture' area with only some very general instructions remaining in the 'design' area.*

Now, in total contrast, take the case of a very different sort of product, the wing spar of, let us say, a jet fighter. In order to combine maximum strength and integrity with minimum weight, the member will almost certainly be machined out of a solid metal beam. By the time the machining is complete as much as 90 per cent of the metal may have been cut away to leave an immensely intricate and precisely calculated shape. The people who designed the spar will not only have refined its shape until it is as close to perfection as possible, they will also have specified the manufacturing process in every last detail.

In fact, the most important piece of information to pass from design to manufacture is not a set of drawings, but a set of numerical control part programs which specify the tools to be used, the paths they must follow, the cutting speeds and feed rates that must be set, and so on. In order to ensure that those programs are correct in every respect and that all 'bugs' have been eliminated from them, the entire manufacturing process will have been simulated on computers so that, by the time the first metal is cut for the first spar, there is no room whatsoever for those involved in manufacturing to exercise any discretion.

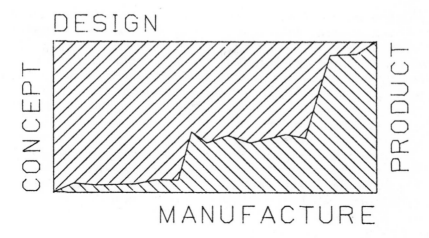

Figure 23. *The use of techniques such as numerical control allows a designer to exercise almost total control over the manufacturing process and, as here, manufacture may involve little more than the implementation of a program that has been refined and tested in every detail.*

In this case, as Figure 23 shows, manufacture has shrunk back into one small corner of the continuum which is dominated by the design function.

The whole thrust of technological development over the past two centuries or so has, clearly, been towards enlarging the design proportion of the continuum in this way and diminishing the manufacture one. This change has been expressed in a whole variety of ways, the most obvious of them being the rise of the professional engineer and designer and the oft-lamented demise of the individual craftsman.

In part, the change has become inevitable as products have become more complicated and have demanded greater and greater precision in their manufacture. In part, it has been due to the invention of mass production, which depends upon components being uniform and inter-changeable. Over the last 40 years or so, it has also been vastly accelerated by the introduction of automation of one sort or another. For once manufacture becomes an automatic process, that is, one which is control-led by machines following preprogrammed instructions, it becomes little more than the 'playing back' of a program written by the designer who will have been obliged to foresee and provide the instructions required at every stage in the process.

In a highly automated plant, the people who mind the machines are no more 'making things' than a person who puts a disc on the record-player is 'making' music. Both are simply supervising machines which are following instructions finished by other people in other places. Today, naturally enough, most such instructions take the form of computer programs, and CAM is in essence concerned with the kind of machines which can play back programs generated by, or with the help of, CAD technology. In a nutshell, just as CAD allows a designer to create simulated objects, and then to simulate their behaviour in order to analyse it, so it also allows for the processes involved in manufacturing those objects to be simulated. The programs which are generated in the course of simulated manufacture can subsequently, if played back on different kinds of machine, be used to control the actual manufacture of actual objects.

As we have explained above, most CAD systems are still restricted to the use of mainly geometrical data – they can understand the shape of objects in great detail, but can know little of how hard or soft, hot or cold, heavy or light, strong or weak, brittle or pliable those objects will be. The term CAM is, similarly, usually applied only to manufacturing processes where the basic control data is geometrical. This means that there are many applications in which computers are used to 'aid' manufacture which, none the less, are not categorized as CAM – the computer control of a chemical process plant, for example, or of injection moulding machines.

Numerical Control: the Basis of CAM

The idea of encoding geometrical instructions in numerical form and using them to control the movements of automated tools was widely practised for several years before computers became involved. When the same machining operation has to be repeated over and over again, there is, obviously, a very strong economic argument for eliminating the need for skilled and highly paid machine operators constantly to tend the tools by substituting some form of automatic control. When the operation is a metal-cutting one – drilling holes, milling shapes out of a solid, turning objects on a lathe – the control function is essentially concerned with moving a cutting tool precisely in two, or sometimes three, dimensions. By the early 1950s it had proved possible to design control units which achieved this extremely successfully. The basic technique was for a set of instructions, a part program, to be written in one of a number of specialized languages developed for the purpose and then to be encoded on punched paper tape. The tape was then fed into the control unit which 'read' it, section by section, and moved a cutting tool accordingly – rather as a pianola implements the instructions encoded on a roll of punched paper.

The usefulness of this basic form of NC was, however, limited. Once a part program had been perfected it did indeed allow a machining operation to be repeated over and over again accurately and (but for a tendency for the tapes to wear with repeated use) reliably. The problem was that actually producing even quite a simple part program was an extremely long and difficult job. A specialist part programmer had to take the designer's drawing and translate all the geometrical forms into one of the NC languages which described the movements the tool was to make and the resulting program then had to be converted into a punched tape.

Inevitably, given that the part programmer was forced to work 'blind' as it were and yet had to take account of many different factors such as the offset for each tool, the position of fixing clamps, the differing cutting speeds and feed rates for various combinations of tool and material, and so forth, problems would be found when the tape was first run. These then had to be analysed, the program corrected and a new tape punched – a cycle that might have to be repeated many times. Obviously, such a lengthy and expensive process of trial and error could not be justified when all that was needed was a small batch of components and even less so when it was a case, (as routinely happens in many a toolroom for example) of producing items as one-offs.

Figure 24 shows the complete sequence of processes involved in NC, and the way in which they translate the geometry created by a designer into the movements of a cutting tool. In theory, by allowing the designer to exercise a sort of remote control over the manufacturing process, NC

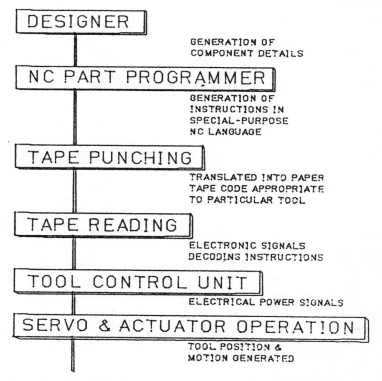

Figure 24. *The sequence of operations involved in orthodox NC – that is, NC that does not involve the use of computers.*

reduced the manufacture section of the continuum to a purely 'mechanical' operation and gave the designer the power to specify both what was made and how it was made in every particular. In practice, however, it was rarely possible before the advent of CAD/CAM for designers to take full advantage of the technique.

The difficulties mostly arose at the first two stages of the process, those that involved the writing of a part program and transferring it on to punched tape. First, (as we have already seen,) the fact that both these operations required much painstaking work by skilled personnel, created a bottleneck that was both physical and economic – there were only so many part programmers and punch operators available in any company, they could do only so much work and, because of the time it took, that work cost a great deal of money. The use of NC could only be justified, therefore, if the cost of creating a part program could be spread over a long production run or if the job required a degree of precision which could not be achieved by any other means.

But, to come on to the second of the difficulties, the degree to which the designer could actually control the manufacturing processes was limited by the fact that the geometrical information created on the drawing board had to pass through the hands of two human intermediaries, each of whom translated it into a different form. Moreover, the first of these intermediaries, the part programmer, had to be given some freedom of judgement and even the power to interpret the designer's wishes to some extent, for programming languages (and tool control systems too, for that matter) can only approximate the ideal tool path with a greater or lesser degree of accuracy, they can never define or follow it with total precision.

Thus, rather than actually putting the designer's hand on the tool controls, as it were, NC had removed the limited amount of discretion and autonomy which the machine operator had previously enjoyed, only to hand it over to the part programmer who, however skilled, was handicapped by the limitations of the techniques at his disposal.

Even though CAM is normally associated with the two alternative systems for imposing computer control upon a machine tool, computer numerical control (CNC) and direct numerical control (DNC), which we shall consider in a moment, the extension of CAD systems to provide computer-assisted part programming is at least as important a development. For, without this bridge linking CAD to CAM, NC, whether exercised by a computer or not, would still be hamstrung by the difficulties outlined above.

Computer-assisted Part Programming

Fortunately, once a designer has created a geometrical model inside a CAD system, that system has most of the information needed to generate the tool path and turn it into a program that will cause a machine tool to produce a real-life object corresponding to the geometrical model. One advantage is that the CAD system, unlike the human part programmer, does not have to analyse the geometrical properties of the 'drawing' which appears on its display screen: on the contrary, it is only because it already 'knows' all there is to know about these properties that it can create the image.)

The system does, of course, have to be provided with some basic information about the tools that will be used. It must, for example, know what offset to allow for (offset is the distance which separates the centre-line of the tool path from the surface of the workpiece – the radius of the tool, in other words), and the movements which the tool may 'legitimately' make, and it may also be able to register the position of clamps that will hold the workpiece in position, or other obstructions.

In most cases the CAD computer will also have to cope with the chore of postprocessing. (This is equivalent to the task performed, in an orthodox NC operation, by the punch operator who must take the part program and, while punching it on to tape, also translate it into a form which is intelligible to the control unit of the machine tool concerned. Although there are no more than a handful of NC part-programming languages in use, of which automated programming tool or APT is the commonest, every control unit will use a slightly different dialect. The unit controlling a simple drilling machine, for example, will accept only very basic instructions, whereas a machining centre with automatic tool-changing equipment will incorporate a controller that is far more versatile and capable of acting on a wide range of instructions covering many operations. Again, obviously, lathes and flame cutting tools, robots and punches, all of which can be controlled by NC programs, will not function very effectively if they are all restricted to a single common language.) Moreover, the control units produced by different manufacturers each have their own idiosyncracies, each demanding the use of a different postprocessing program. Postprocessing, then, is a vital function and a very time-consuming one, even for a computer. But it is, essentially, mechanical and therefore poses no real difficulties for the machine other than the demands which it makes upon processing time.

The final postprocessed part program may then be output via a tape punch, today still the commonest method of communicating it to the machines which control the manufacturing phase but losing ground to magnetic tape or disc which can be loaded into another computer; in other instances the part program may be transferred directly from the graphics computer of the CAD system to its counterpart in the CAM system; or, in some cases, the graphics computer may even run the program itself and exercise direct control over a tool.

The advantages of using the CAD computer to write part programs become obvious (if you look at Figure 25 and compare it with Figure 24, which showed the sequence of operations involved in orthodox NC. The most labour intensive, skilled and cumbersome tasks have now all been handed over to the machine and, equally importantly, the original geometrical information created by the designer keeps its integrity – rather than being passed from hand to hand as the programming progresses, the information stays in one place and is translated from one form to another by programs that are reliable and predictable. It is, of course, necessary for the computer's part-programming work to be supervised by a skilled human part programmer – apart from anything else there will always be decisions about the best tools to use, the choice of cutting speed and feed rate, etc. But the task of the human programmer is vastly eased by the computer's ability to 'show him what it is doing'.

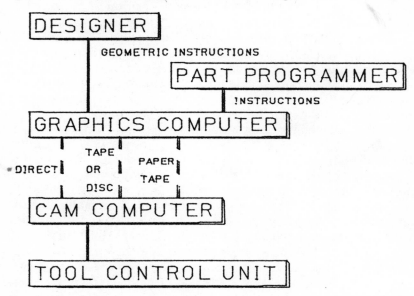

Figure 25. *Computer-assisted part programming. Not only is the sequence of operations far simpler and shorter than that shown in Figure 24, but, crucially, the graphics computer itself has taken over the most error-prone and time-consuming phases of the process.*

Interpreting a part program and checking it for accuracy no longer involves examining a long sequence of coded instructions in a language intelligible only to experts; the CAD system which can show you what an object will look like can equally well show you what a tool path will look like. Figure 26, for example, shows the computer-generated tool path required to cut some of the surfaces of the object which we used as an example of a geometrical model in the previous chapter.

The ability to see a tool path – in some cases, a system will actually simulate the movements of the tool in three dimensions in 'real time' –

is enormously advantageous. If, for example, the part programmer is worried that there will be insufficient tolerance at some point or if he has to choose between two possible approximations of a shape, the problem can be inspected visually and, maybe, even put up on the designer's screen so that he can take the final decision. In many cases, it is possible to experiment with different positions for the clamps or to program in movements necessary to avoid other obstructions rather than having to sort such problems out on the factory floor where precious machine time will be wasted.

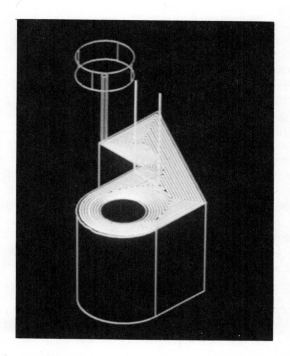

Figure 26. *A computer-generated tool path showing the tool movements required to machine the upper surface of the model used earlier to illustrate the discussion of CAD. (Figure reproduced by courtesy of Computervision.)*

Perhaps the best way of looking at computer-assisted part programming is as yet another example of the computer's ability to simulate reality. In this case what are being simulated are the actual processes of manufacture. You could say, in fact, that the CAD system provides a stage upon which the designer can rehearse the business of actually making his brainchild. Like all rehearsals, this affords an opportunity to see when things are not working and why, and to spot problems that might otherwise have gone unrecognized until the performance actually began.

So far, admittedly, NC is applied to a rather limited range of processes:

those involving the traditional machine shop tools such as lathes, milling machines etc and other metal-cutting techniques such as punching and flame cutting. But, as we shall see shortly when we come to discuss flexible and computer-integrated manufacture, the repertoire is growing. Already NC techniques are being applied to the control of assembly systems and a link-up has been established with the sister subject of robotics. The optimists are even suggesting that it will not be long before we are in a position to simulate and preprogram the operations of an entire factory 'off-line' on what will essentially be a CAD system. Already, indeed, CAD is being used to design and simulate the operation of very substantial parts of a factory, such as robot welding lines.

Direct Numerical Control

Confusingly, direct numerical control is just as much a system for controlling NC operations with a computer as its rival computer numerical control. The difference between the two systems is not unlike that which distinguishes a stand-alone CAD workstation from a system built around a central computer (see next chapter). For, whereas CNC employs microcomputers on a one-micro-per-tool basis, DNC, which was the first of the two to appear on the scene, is a centralized system. DNC was a product of the early 1970s when, although the minicomputer was already beginning to topple the mainframe from its pedestal, it was not yet clear that the cost and scale of the hardware would continue to shrink as it did.

As can be seen from Figure 27, the structure of a DNC system is highly centralized. In its most basic form, DNC is achieved by simply removing the tape readers from a number of NC machine tools and connecting the control units to a single minicomputer which, together with its associated storage devices, acts as the central control and memory for the entire installation. Programs may be loaded into the mini either via a tape reader or directly in software form. In either case the most vulnerable and unreliable part of the old system is eliminated, for, even if tape is used, it now only has to be read once rather than being rerun for each cycle of operations, a process that rapidly led to wear and tear and, thus, to mistakes.

This is an advantage which is shared by CNC, as is the facility for inputting instructions directly into the controlling computer in order to modify, improve or correct a program that has already been loaded. As has already been explained, computer-assisted part programming on the CAD system allows the programmer actually to see the tool path on the screen, and thus to identify and eliminate many of the bugs which a program may contain before it reaches the factory floor. But the fact that a program can be modified while in computer memory at the CAM stage further improves matters since it is no longer necessary, when a fault is discovered, to take the program 'off-line' as it were, rewrite the section involved, repunch the tape and start all over again. Now, most problems can be put right as and when they emerge with the minimum waste of machine time.

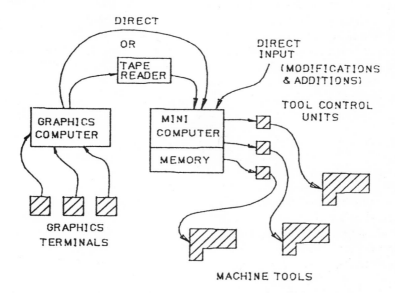

Figure 27. *The relationship and flow of information between the elements of a CAD system and a DNC installation.*

In the case of most DNC systems the part program is taken off the graphics computer in its initial form and the DNC computer itself performs the postprocessing function. This has two main advantages: it frees the graphics computer from a task which would otherwise be a substantial call upon its time, and it allows programs to be stored in a form which is not specific to any particular system or machine tool.

The most important single benefit of DNC, however, derives from the fact that it allows a whole repertoire of programs to be stored in instantly accessible form and the system can, therefore, switch any of the machine tools under its control from one program to another virtually on demand provided tools and programs are available. This makes DNC the first choice for anyone who is contemplating systems such as flexible manufacturing in which a number of tools must work together and adapt rapidly from one sequence of operations to another.

The main drawback of DNC is, of course, its all or nothing nature. It allows no opportunity for newcomers to test the water, so to speak, but requires that they plunge straight into a major investment which, if it goes badly wrong, may put entire sections of a business out of action.

Computer Numerical Control

CNC, in contrast, allows computer control to be introduced to the machine shop in piecemeal fashion, for the most basic CNC tool is simply one in which the 'hard-wired' control unit with its electro-mechanical relays has been replaced by a microcomputer. Input is normally, though not always, via a tape reader and there is, as Figure 28 shows, provision for direct input of instructions in order to correct or modify a program once it has been loaded.

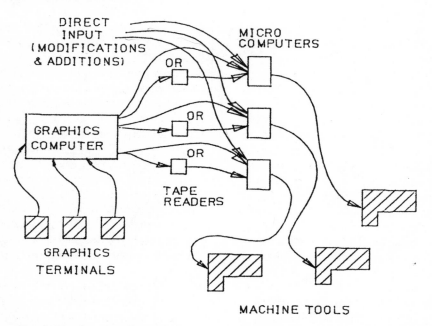

Figure 28. *The relationship and the flow of information between the elements of a CAD system and a CNC installation.*

The principal disadvantage of this arrangement is that the microcomputer will normally be able to store only one part program at a time, and so a new program must be loaded every time a tool is switched from one task to another. The limited power of the micro also means that the postprocessing function must be handled by the graphics computer.

The Future of Numerical Control

It is unlikely, however, that an investment in CAM technology will involve the sort of stark choices which these descriptions of DNC and CNC suggest for very much longer. There is indeed already ample evidence that all sorts of hybridizations between the two systems will occur in the relatively near future and, as a consequence, the distinctions are likely to become murky.

In most cases, for example, new DNC systems use micro-based control units rather than the old-fashioned hard-wired type, thus becoming, in all but name, a series of CNC tools with an additional level of supervisory control and program storage added. And while DNC systems grow down to meet CNC ones, the reverse is also happening in that many CNC tools are now designed to be used as either stand-alone units or parts of some larger system.

Such hybrid systems will combine the best features of both DNC and CNC. A micro-based control unit, for instance, is inherently more flexible than the hard-wired variety. Moreover, even if the micros are working under the supervision of a larger host computer they will make fewer demands on the host machine because an entire program can be loaded into their memories rather than having to be repeatedly fed in line by line, as with a hard-wired control unit.

It will also be possible to start with a few individual CNC machine tools, or even a single one, and then gradually build up the installation adding further levels of control as and when they are required. Already systems with more complex hierarchies of control are appearing on the scene. Some DNC systems incorporate an intermediate level with a number of the new super-micros each controlling a few tools under the overall supervision of a minicomputer, while the expansion of DNC into what are called computer-integrated manufacturing (CIM) systems (see below) involves grouping a series of DNC systems together under mainframe control.

Ultimately, of course, the aim must be a totally automated factory with many different levels of computer control and human beings simply feeding in a description of what they require via some 'super-CAD' terminal and waiting for a finished product to emerge at the other end. But the signs are that such ideas, even if theoretically feasible, are for some time to come likely to be confined to the realms of fantasy and, perhaps, a few showpiece installations. In the real world, industrialists are happier replacing and modernizing existing plants piecemeal rather than committing themselves to the construction of futuristic factories costing hundreds of millions of dollars.

What are already beginning to appear in industry are various kinds of compromise systems which allow some of the advantages of greater automation to be realized without involving the sort of total commitment to 'computer-aided-everything' which the high priests of CAD/CAM might like to dream about.

Flexible Manufacturing Systems

A flexible manufacturing system (FMS) is a concept which is, in itself, so flexible that it almost defies definition, except in terms of the purpose it is intended to fulfil. Basically, the idea is that between the large-scale industries which can both afford and make proper use of 'hard automation' such as that involved in CIM (see below), and small machine shops containing perhaps two or three NC or CNC tools where further automation would be inappropriate, lies a no man's land of medium-sized enterprises devoted to batch production. Typically, the operations of such a business are on a sufficiently large scale and of sufficient complexity for automation to be cost effective, yet do not involve the long production runs which could justify the installation of special purpose equipment.

What is required is a system which is integrated, in that it provides for the automatic transfer of workpieces from one machine station to another, but is also flexible in the sense that it can be switched from one task to another.

CNC and DNC, or some combination of the two, can be used to provide the individual tools with suitable repertoires of programs which can be altered or widened as the need arises. But different tasks may require different tools to be used in different combinations or sequences, with different checks being applied at different stages. The industrial robot provides at least a partial answer to this problem. Robots are relatively easily equipped with 'hands' or end effectors for loading and unloading machine tools; being themselves computer controlled, they can readily be interfaced with the machine tools' control units in order to coordinate their movements; and a modern robot can store a variety of different programs.

An FMS is therefore a 'cell' made up of two or more machine tools plus, perhaps, testing and calibration devices, grouped around a robot which services the tools. A typical arrangement is shown in Figure 29. At present the self-sufficiency of such a 'cell' is limited by the fact that the robot's senses, if any, are extremely rudimentary – it will therefore be reliant on a human operative, or on hard automation such as conveyors and indexing machines, to deliver workpieces in predetermined and consistent positions and orientations. Human help and intervention also become necessary if the manufacturing process requires work to be transferred between cells.

The main feature of flexible manufacturing is the fact that the individual machine, the machining centres, lathes, robots, etc, which make up a cell retain a high degree of autonomy. Essentially, each is separately programmed and communication between them is limited to ensuring that their actions are coordinated. Because the system is subject to no overall supervisory control, the task of reorganizing it for a novel sequence of processes can be a lengthy one. Although, in theory, the information

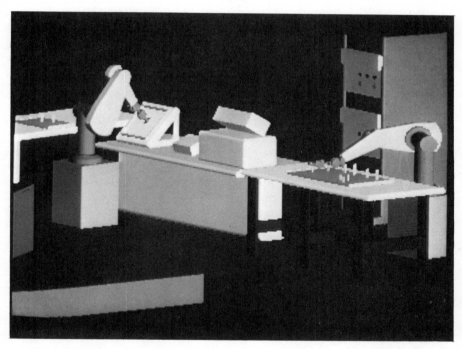

Figure 29. *The layout of a typical flexible manufacturing system showing two robots loading machine tools from prepositional racks of piece-parts. (Reproduced by courtesy of Computervision.)*

required to accomplish it may be generated by CAD techniques, there is a long way to go before it can be conveyed directly to the machine's own control units without the need for the services of human programmers.

Computer-integrated Manufacturing

The next step up, so to speak, from flexible manufacturing is computer-integrated manufacturing (CIM) (or engineering (CIE) – the two are in practice synonymous). In this case tools are assembled in much larger combinations and linked by means of hard automation involving such things as transfer lines, pallet conveyors, automatic trolley systems, etc. Rather than being semi-autonomous and separately programmed all the elements in the system are integrated under a centralized and hierarchical control system. A CIM system is also likely to be tied into data bases and information systems concerned with functions such as inventory control, production planning, etc, unlike an FMS cell which will have access only to geometrical information.

Typically, a CIM system might be based on a mainframe or a powerful minicomputer which acts as a program 'bank' for all the constituent machines and exercises its overall authority via a network of satellite computers which are, in turn, linked to the individual machine control units. The programs can thus be taken out of the central memory and loaded into control units as required while the central computer and its satellites retain control over the flow of work through the system as a whole.

Despite the theoretical attractions of CIM it must be said that, except in the case of a very few showpieces, they have yet to be realized in practice. With the price tag measured in hundreds of millions, and with the technology still in a state of rapid flux, there is little incentive for any but the largest business to commit itself to such a radical development. It seems likely that CIM, in its purest form, will remain an ideal that will be unrealized for some time to come for all but very special processes, which is not to say that many of the ideas involved will not be implemented on a piecemeal basis. The distinction between CIM and FMS, for example, is in practice not anything like so clear cut as it might appear in theory. As hard automation becomes more flexible, and robots become more mobile and more highly skilled, the two may merge – on the factory floor if not in the textbooks.

Group Technology

The gradual merging of the different systems of numerical control which we have discussed, together with developments such as FMS and CIM, can be thought of as more or less 'pure' CAM. They are largely independent of parallel advances in the CAD field which they may or may not make use of. But the newly fashionable area of group technology is centred in the no man's land between CAD and CAM, taking advantage of the increasing sophistication in the former field in order to speed the development of the latter.

Traditionally, engineers have classified their products according to their functions – a crank is a crank whether it is two inches long and made of aluminium alloy or two feet long and made of top quality steel, an automobile wheel is a wheel no matter whether it is a product of the press shop costing no more than a pound or two to make or the result of a series of machining operations with a price-tag many times higher. How the component is made and what it is made of has been of only secondary importance compared with what it is made for.

Logically, however, a manufacturer should attach more importance to the processes that go into the manufacture of an object than to its ultimate function. For the shape of the thing, the sequence and nature of the operations needed to make it and the materials of which it is made are the factors that determine where it is made, by whom and at what cost.

Clearly, the extent to which different objects share similar characteristics will also reflect the extent to which their manufacture involves similar tools, skills and processes.

The key idea in group technology is that objects which might, in traditional terms, be viewed as having nothing at all in common may in fact share features which are extremely significant from the point of view of the manufacturer. If, to take an analogy from everyday life, making a cheese soufflé and making a chocolate soufflé require the same basic skills, then the fact that one will be served at the start of a meal and the other as a dessert is, so far as the cook is concerned, a secondary consideration. Or, to take an industrial example, to say that bedframes and bicycles are totally dissimilar products, sold in different places to different people for different purposes might be to miss the point that both can be made by welding lengths of steel tubing together.

The geometrical properties of an object are likely to be among its most significant characteristics from the manufacturing point of view – similar shapes are likely, all other things being equal, to involve the use of the same processes and the same tools. Figure 30 shows two groups of objects, which, while heterogeneous to the uninformed eye, are closely related in terms of their fundamental geometries and/or the manufacturing processes required to make them; on the other hand, the groups of objects in Figure 31 have superficially similar geometries but are made by totally dissimilar techniques.

The purpose of group technology is to classify objects in a fashion which reflects their 'true' relationship to each other, so that things which are a similar shape and made in similar ways are 'closely related' whereas things which are made in different ways are far apart from one another even if they look the same.

There are two main purposes which group technology is expected to serve. First, it may save much wasted time in the design department since whenever a new requirement is specified it will be possible to search the existing range of designs to see whether it does not contain an item which, with or without some modification, might meet this new need. Obviously, a CAD system with a memory that is, potentially at least, an electronic catalogue of geometrical models could be adapted to provide the ideal foundation for a group technology database.

Second, group technology should be indispensable in organizing the efficient functioning of FMS and CIM installations since it provides exactly the sort of data which a computer will need to plan the proper utilization of the equipment at its disposal. If, for example, two different products both involve the use of the same set of part programs at one stage in their manufacture then it may make sense, once the operations have been completed on a batch of the first product, to follow on with a batch of the second even if they are then put aside for completion on some other occasion. Similarly, if one task involves the use of only some

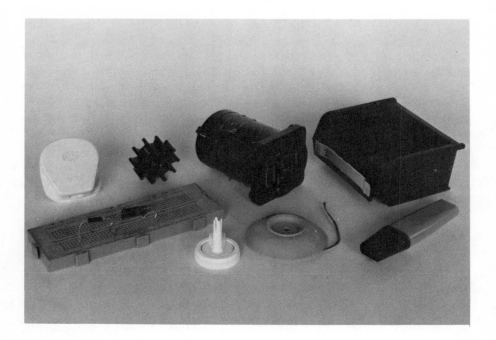

Figure 30. *Although superficially dissimilar from one another all these objects are made of the same material – plastic – and produced by moulding processes. They would thus be considered 'closely related' in terms of group technology.*

of the machines in a cell, then it may make sense to search the schedule of future jobs to find one which will make use of those machines which would otherwise be idle.

Already the theorists of group technology are busy, on paper at least, reorganizing the layout of factories so that instead of machine tools being grouped by category and function – all the turret lathes in one bay, all the drills in another, etc – they are assembled in lines or cells each of which is designed to carry out the entire sequence of operations which the manu-facture of one group of products requires.

Again, it is necessary to stress that such futuristic, idealized set-ups are unlikely to be built in any great numbers. But that is not to say that the principles of group technology will not be widely employed on a smaller, more piecemeal scale. This is likely to become more and more the case as

Figure 31. *This group of objects all share the same cylindrical form; but in manufacturing terms they are totally dissimilar, being made of different materials and produced by a whole variety of processes from glass-blowing to metal removal.*

the sort of developments in CAD which we shall discuss in Chapter 6 make it possible for a computer system to understand properties of objects other than those based solely on their geometry.

Summary: from CAD/CAM to CADAM

By this point, we hope, it will have become clear that in our view the various and often diverse developments that are commonly grouped together under the CAD/CAM label do share one central thrust: they are all directed, in one way or another, at transferring control over the manufacturing process into the hands of those who design the products. They are, in other words, stages along a road that leads to a situation in which the manufacture of an object will involve no more than the automatic 'playing' back of a program which has already been completed, tested and perfected by simulation before it is ever put into effect. The designer, and those other people associated with the preprogramming of the manufacturing process such as part programmers, group technologists, etc, will be in a position rather like that of the contemporary electronic composer who, instead of writing a musical score which must then be interpreted by a conductor and played by an orchestra, can 'play' all the instrumental parts on a synthesizer and mix them in order to produce a definitive record of his own composition which can then be played, with absolutely guaranteed and consistent results, on any record player, any time, anywhere.

The fact that it is now fashionable to substitute the smoother and more coherent-sounding acronym CADAM* for CAD/CAM is perhaps symptomatic of a coming together that will, ultimately, result in a single unified technology embracing every stage in the design-manufacture continuum from the designer's first rough sketch to the final packaging of the product, and be applicable in fields from civil engineering to integrated circuits. In the meantime, however, the would-be user would do well to remember that contemporary CAD/CAM is still a diverse and often fragmented field fraught with mechanical incompatibilities and human misapprehensions.

*Although CADAM is used almost interchangeably with CAD/CAM in Britain and elsewhere, this is not the case in the US where the word is a registered trademark of the Lockheed Corporation and of IBM who have adopted the term for some of their software.

Chapter 3
The Elements of a CAD System

The key component in any CAD system is, by definition, a computer. The fact that it may also be the one which is superficially the least obtrusive makes it all the more important to remember that all the rest of the paraphernalia which go to make up a complete system play a subsidiary role. The different kinds of display technology and the various combinations of input and output devices, which we will be discussing later in the chapter, are basically no more than a means of ensuring that the interaction between designer and computer should be as easy, efficient and effective as possible.

From Mainframe to Mini

Given that a computer is the *sine qua non* of a CAD system, it follows that the amount of computer power available – the combination of processing speed and memory capacity at the designer's disposal – is the most important factor in determining a system's abilities. No matter how advanced and sophisticated the rest of the equipment, a system with too little computer power will be like an 'under-engined' car, and there will be firm limits to what can be done with it. It should be added, of course, that just as installing a bigger engine does not automatically produce a better car unless other systems are redesigned to make use of the additional horsepower, so adding crude increments of computer power will not necessarily improve a CAD system that is inadequate in other ways.

We have already explained that it was the availability of cheap and reliable silicon-based hardware which transformed CAD into a general-purpose tool. But it would be misleading to suggest that the development of CAD has simply been a matter of making more computer power available at lower cost, though the economic factor has certainly been a vital one. What has happened is rather more complicated than that and has involved two quite distinct developments which are now just beginning to blend into one another. The first of these, associated with the early stages of the silicon revolution which saw the hegemony of the giant mainframe computer overthrown with the arrival of the minicomputer, freed CAD from the need to queue up, along with other applications, for a share in a limited supply of computer power. It meant, in short, that rather than having to 'plug in' to the resources of a mainframe, CAD could afford the services of a 'dedicated' minicomputer. The second stage of development, resulting from the introduction of the microprocessor, has involved the decentralization, or devolution, of computer power *within* a CAD system – it has, in other words, seen the growth of systems containing several separate computers.

In the 1960s, the pioneering days of CAD, computer power was available in only one form, a mainframe of the second or third generation. The philosophy which determined the way in which these machines were designed and marketed was a result of one overriding factor – the comparatively high cost of hardware. While electronic systems were still built up out of individual components, or comparatively small-scale integrated circuits, and while the task of assembling them was labour intensive – as was the case, especially, with the magnetic core memories which were essential to the 1960s' machines – there was little prospect of any significant fall in computer prices.

Virtually all manufacturers therefore adopted the following strategy, which is usually credited to an IBM analyst called Herb Grosch. Since they saw no way of mass-producing computers at mass-market prices, they did not bother to consider the 'Model T market', as it were; instead they concentrated on producing 'Rolls Royces' – even bigger, more powerful, more versatile and, of course, more expensive machines. The idea was that a single one of these giants would supply all the computer power that a large corporation (or a number of businesses taking advantage of the 'time-sharing' facilities of a computer bureau) could possibly require.

In the case of most commercial computer applications, the strategy had much to recommend it. For much of the work involved was simple data processing, and routine 'paperwork' functions such as book keeping, invoicing and inventory control were relatively easy to 'computerize' within the framework of the Grosch strategy. They required very large amounts of information to be stored, but that could be achieved with 'back-up' devices such as discs or tape, and the central processing unit (CPU) of a big mainframe could take data in and out of store and deal with relatively modest computational demands at a rate which allowed it to service hundreds of 'customers', apparently almost simultaneously. We say 'apparently' because it was of course an illusion. Like every other 'serial' or 'von Neumann' computer, a mainframe, however big, has only one central processor and executes only one logical operation at a time. It only *seems* to be doing several things at once because it may be capable of performing hundreds of thousands of operations each second.

CAD, however, makes very heavy demands upon processing power. A single designer, doing routine drafting work, may be taking up more of the central processor's time than a dozen or more users involved in clerical work or data processing. And when transformations must be applied to a complex model, or NC part programs require postprocessing, the CPU's attention must be monopolized for minutes at a time – a virtual eternity! Moreover, CAD programs also require a great deal of random access memory (RAM) capacity, something that was in very short supply before the arrival of fourth-generation computers equipped with silicon memory chips.

To most industries, therefore, CAD appear not as a wonderful new opportunity to make use of their corporate mainframe, but as a greedy, parasitic technology which, if once allowed to plug into the machine, would quickly consume far more than its fair share of the available power. There were, of course, exceptions. Giant corporations, like General Motors whose research laboratories produced the first truly interactive system in the early 1960s, could afford the lavish investments in computer power which CAD demanded. In other cases there was really no choice; the problems of laying out integrated circuits or performing the stress calculations involved in designing high-performance aircraft were beginning to exceed the limits of what was possible using traditional methods of drafting and analysis.

The introduction of the first minicomputers in the early 1970s saw the start of the steady shrinkage in the size and cost of computer hardware which still continues today. As a result, it became possible to provide more and more computer power for CAD purposes without prohibitive cost problems. The fact that, with the arrival of the mini, at least one zero was knocked off of the price-tag of 'a computer' made it feasible to design systems which had the exclusive use of a 'dedicated' machine. Admittedly, the first minis did not have the same power as contemporary mainframes – though the gap between the two categories has been steadily narrowing for a decade or more now – but, freed of the need to service other customers, a mini of the mid 1970s could cope with a system containing perhaps half a dozen workstations without any problems at all.

The arrival of minicomputer-based systems also resulted in the birth of what has, today, become a separate and quite substantial CAD/CAM industry. For whereas most research and development in the mainframe era had been conducted 'in-house' by established businesses, largely with a view to meeting their own particular needs, it was now possible to envisage the widespread use of CAD as a general-purpose design tool. A number of new companies quickly sprang into existence to seize the opportunities that were opening up and it is these firms which today provide the vast majority of the 'turnkey' systems on the market. Typically, such systems (an example of which is shown in Figure 32) are built around a standard, medium-sized minicomputer from one of the major manufacturers, with the CAD manufacturer providing much of the peripheral equipment and a wide variety of software, both general purpose and specialized.

The power and flexibility of such systems was greatly widened by the

Figure 32. *A representative example of the wide range of 'turnkey' systems currently on the market, Computervision's CDS 3000 system. On the desk are the display screen, graphics tablet, keyboard and hard copy unit that furnish a single workstation; behind, the CPU and disc drive unit that are shared by the entire system. (Photograph reproduced by courtesy of Computervision.)*

next development, which resulted from the introduction of the microprocessor and the availability of cheap, mass-produced microcomputers.

Enter the Micro – Distributing 'Intelligence'

The introduction of the minicomputer freed CAD from the need to share computer power with other users, but it did not, at first, change the highly centralized fashion in which computer power was deployed. In most cases, a workstation was no more than a console through which a designer and the central computer could communicate with each other. At most, when raster display was used (see page 69), the terminal might contain the minimal amount of computer hardware needed to control its own display screen.

But it quickly became apparent that, first, the availability of microprocessors at virtually negligible cost made it both possible and economical to provide a large number of subsidiary, decentralized 'intelligences' within a single system and, second, that there would be great advantages in this. The point was that as CAD programming became more elaborate the need to channel all information processing through the minicomputer's CPU was beginning to produce a bottleneck, restricting the number of terminals that could be serviced and resulting in 'customers' being kept waiting.

The manufacturers responded to this problem by producing the first 'intelligent terminals'. By building one or more microprocessors and some memory capacity into each terminal, and providing a modest amount of back-up memory in the shape of disc drive units, it was possible to make a terminal sufficiently 'intelligent' to control its own display screen, and to carry out some of the routine graphics functions. The terminal, for example, could handle the transformations involved in zooming in or panning across a given view of a model, though the model itself would still be stored in memory devices associated with the central computer. The power of the 'host' computer could also be called upon whenever more elaborate tasks were involved.

Essentially, in a network of this kind, the 'host' minicomputer is situated at the focus of a number of radiating links, each connecting it to one workstation – the arrangement shown in Figure 33. It could be thought of as a kind of omniscient, always accessible, boss coordinating and controlling a number of subordinates, each of which operates semi-autonomously

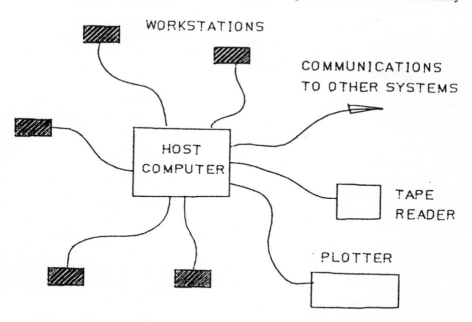

Figure 33. *A host-based CAD system is centred upon a minicomputer which services a number (in practice anything between two and, say, twenty) of workstations. Each workstation will contain its own microcomputer and limited memory capacity, but the power of the host machine will be needed for anything other than routine graphical functions. The minicomputer also drives a plotter serving the entire system and has access to a tape reader, which provides the system's archival storage.*

but can always rely upon the boss's help when a problem turns up which is beyond its own abilities. Because the minicomputer is spared the task of constantly supervising all of the workstations all of the time, it is possible for it to react faster when it is needed and for it to handle programs which might otherwise demand too large a share of its power.

But the microcomputer, especially the later generations of machines with powers that began to creep up the scale towards the bottom end of the mini range, also made possible a quite distinct and different development – the self-contained, 'stand-alone' workstation with its own built-in source of computer power. At the most basic level, a stand-alone system may amount to no more than an off-the-shelf microcomputer – an Apple or an IBM PC, for example – equipped with some specialized graphics software. Such systems are very far from being mere playthings; at Brunel University for example, students are taught advanced design work on Apple, Sirius and BBC micros, using peripheral equipment available off the shelf and software specially developed for the purpose by one of the present authors (AJM).

Further up the scale purpose-built, stand-alone terminals built around one of the 'super-micros' are available from a number of manufacturers. In most cases, provision is made for such systems to be linked together to form a ring. The enormous progress in the design and organization of computer networks which has been made over the past decade allows the individual machines to cooperate in an extremely sophisticated and flexible fashion. Not only can information be exchanged between one terminal and another, but resources can be shared, models can be created separately and then assembled to form a larger model, and so forth. The basic 'architecture' of such a system, shown in Figure 34, can be contrasted with the very different arrangement shown in Figure 33.

This contrast dramatizes the dilemma which has, until fairly recently, faced anyone who contemplated the purchase of a medium-sized CAD system. If all went well, they might expect to be using CAD on a scale which would justify the purchase of a system of the first and more expensive type based upon a central minicomputer with satellite workstations. But, although it is usually possible to start with a few workstations and add more as experience and demand increase, the most expensive single piece of hardware is likely to be the minicomputer and, clearly, it would be folly to start with one that could not supply all the computer power that the system would demand when it reached its full size.

Opting for a system of this kind, therefore, requires a very substantial initial investment in a machine that may not be used to its full capacity for some time to come. Perhaps even more seriously, it also involves the purchaser in a more or less firm commitment to one manufacturer – something which, as we shall show in Chapter 9, may subsequently be a cause for regret.

A network of stand-alone workstations, on the other hand, is an

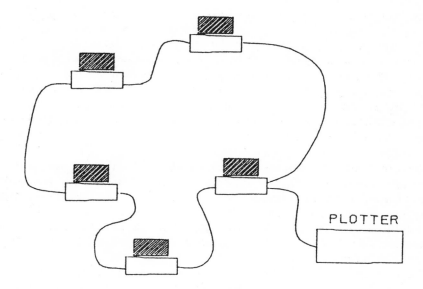

Figure 34. *A series of self-contained, stand-alone workstations, each containing its own microcomputer and memory facilities, can be linked together to form a ring, thus offering the possibility of building a system up in stages. Although such a system may lack the overall power of one built around a central minicomputer, it will none the less meet the needs of many users. Normally, in designing ring structures, no provision is made for direct access to archival storage.*

attractive option in many cases simply because it makes it possible to start with the minimum commitment – the purchase of a single workstation. By adding further workstations this toe-in-the-water situation will gradually be changed into one of total immersion in CAD as and when results justify it or demand requires it. Unfortunately, though, a network of interlinked micros will never be able to match the power of a minicomputer, nor will it be able to handle anything like the equivalent amount of back-up memory, though this shortcoming can be remedied, at some cost in terms of speed and efficiency, by incorporating a link or 'node' in the ring which is dedicated to storage. In cases where a system is ultimately expected to cope with problems that are computationally demanding, or where a substantial memory capacity is important, caution may prove to be a fatal error.

Happily, however, there is every indication that the two kinds of system will shortly meet in the middle! In the case of a centralized minicomputer-based system there are many advantages in allowing the terminals to pass messages to and from one another, as well as to and from the central computer. Apart from anything else, this spares the CPU of the host computer the administrative chores involved in acting as a central exchange. At the same time, workstations designed to stand alone or to form part of a network are increasingly being equipped with the wherewithal to make use of the services of a more powerful computer as well.

Thus it seems that the purchaser will soon no longer be faced with having to make a choice between systems with radically different structures. Instead all systems will, when fully developed, contain both a large 'service' computer and a network of interlinked workstations, with communication links joining all the local intelligences to each other and to the powerful resource of the larger machine – the arrangement shown in Figure 35.

In the future, it will not be a question of what structure you are ultimately aiming for, but rather of how you start building it and when and in what order the various elements are added. This will certainly be a great advance in flexibility, and make the prospect of switching to CAD a considerably less daunting one. But the full advantages of this sort of development will only be reaped when the adoption of industry-wide standards

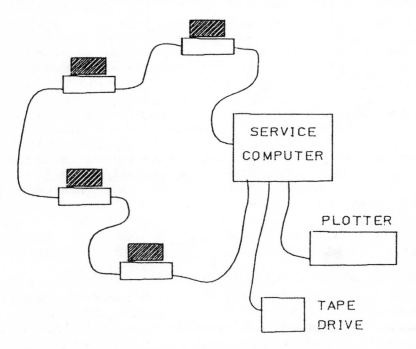

Figure 35. *This is likely to be the sort of structure used in the majority of CAD systems in the future. It retains the flexibility of the network of self-contained micro-based workstations shown in Figure 34, but by building a larger service computer into the ring it also offers the sheer number-crunching power of the host-based structure shown in Figure 33. The minicomputer, with its greater resources, can also relieve the micros of chores such as controlling the plotter, and it allows the system to have access to a tape reader for archival purposes.*

allows for a far greater degree of interchangeability between different brands of hardware than is at present the case.

The development of networks of this kind will also require more work to be done on the problems involved in controlling the flow of data within a complex system. In a centralized structure the host computer retains control over all the information in the system, but once a networked design is adopted difficulties begin to arise – who authorizes the use of data? Who changes it? Who owns it? Who controls access to it? These are issues which will become more pressing as networks grow larger.

The purchaser who does not wish to incur the higher costs and additional hassle involved in buying an 'assembled' system (a system built up out of disparate elements from various sources, as opposed to a 'ready-made' or 'turnkey' system, see page 165) will find that the choice of computer is just one factor which must be taken into account when selecting a system. Essentially, the manufacturer of a turnkey system will have chosen the computer (or, possibly, the choice of computers) which he feels best matches the rest of the package which he is offering. Within rather narrow limits, the customer will have to decide to take the complete system, the good with the bad, or opt for an alternative one.

A word of caution may, however, be in order given the current enthusiasm for stringing numbers of general-purpose micros together to form a ring or Local Area Network (LAN). In CAD applications it is unlikely that any development of this technology will produce systems that match the ability of one incorporating a larger host or service computer. It is also essential to distinguish between standard micro's such as the IBM PC-XT and compatible machines and purpose-built CAD workstations from the same manufacturers; even though the latter may be based upon standard hardware they incorporate a good many 'extras' specifically designed for the CAD role and will, therefore, have great advantages over the standard machine.

Memory and Storage Devices

In any computer system provision must be made for the storage of two quite distinct categories of information: programs and data. Ideally, the computer's CPU would have instantaneous access to every single program it required and to all the data it was currently concerned with – rather as the ideal human worker would be able to remember every single skill and every single fact his or her job required and would never have to look up a procedure or check a piece of data.

In practice, of course, it is as impossible to design such perfection into computer hardware as it is to instil it in human flesh and blood. Compromises have to be made and, in the case of computers, the benefits of instant recall must be traded off against the lower price and greater capacity of less accessible storage. The general rule that governs all comparisons between the various kinds of memory and storage devices is that fast means small and expensive while slow means large and cheap. All the terms are, naturally, relative. Today, a 'small' memory may be able to store millions of bits of information and the cost per bit may be measured in tiny fractions of a penny, and information in a 'slow' storage device may, nevertheless, be accessible in a few thousandths of a second.

But this does not mean that the distinctions are academic and unimportant. In CAD applications, where the execution of a single instruction may require millions of bits of information to be retrieved from memory, processed and replaced in memory, a difference in access speeds which sounds trivial may well make all the difference between a system which, from the designer's point of view, is responsive and enjoyable to work with and one which is lethargic and frustrating.

When you buy a piece of 'software' today, whatever its purpose, what you actually get is a disc of one sort or another – in the case of most CAD systems a hard disc. But how the programs are stored within the system itself will depend on their nature. There are some programs which, in any CAD system, will always have to be immediately accessible – the obvious examples being those needed to control the display screen and the graphics programs which are used actually to construct a model – and which will be lodged in the computer's electronic core stage. Others, such as the programming required to achieve routine transformations of the model on display, must be on call, as it were, but will not be needed from moment to moment. These will be in rapidly accessible back-up devices, normally disc drive units; yet others, specialized analytical programs perhaps, or the postprocessing software associated with particular machine tools, may be needed only on occasion and may be loaded on to the disc drive as and when they are required. Different kinds of programs will, therefore, be stored in different media and, very probably, at different locations within the system.

Data, too, can be classified according to the speed and regularity with which it will be necessary to refer to it and assigned to appropriate types of storage. For the moment, at least, virtually every CAD system files all data according to the model it forms part of or is associated with. (As we shall see in Chapter 5 this is something that may well change as developments like 'expert systems' allow more general knowledge to be accumulated in a system.) But a large model, a complex assembly perhaps, may contain many separate sub-models, and a variety of different sorts of data, only some of which may be relevant at any particular time. It must also be remembered that the particular views or representations of a model which will be displayed on the screen, though based upon the model, are also quite distinct from it. It is perfectly possible, therefore, that a model in its entirety may be stored on disc, while a part of it is held in core storage and a part of that part, the view actually being shown, is in the buffer memory which controls the display.

The most accessible place for any information to be stored is, of course, the computer's own electronic core memory. It could be said that the machine 'knows' the information held in core storage in rather the same sense that a person knows something which they have memorized. (The fact that 'memory' and 'storage' are used almost interchangeably in this field is just a symptom of the confusion which surrounds it.) Because any information which the computer is to process must pass through this internal memory, if only transitorily, its capacity places some absolute limit on the system's abilities – this is why the amount of RAM is used as a yardstick for measuring the capacity of a microcomputer.

The amount of RAM available in general-purpose micros, such as, say, the Apple or the Commodore ranges, does restrict the extent to which they can be used for CAD purposes. There would not, for example, be room in their memory for the more elaborate analytical programs, even if the user had the patience to wait for them to be executed. Nor could they store and transform large models, even if their display technology was capable of handling them. But in most purpose-built CAD systems the amount of internal memory will be such that it affects only the performance of the system rather than its absolute ability. The fact, for instance, that one program may have to be completed and downloaded from internal memory before another can be loaded and run, may make the system slow and tedious to work with but does not, in principle, debar it from running those programs.

In systems where processing power has been devolved to individual workstations, with or without a service or host computer to lend a hand when needed, internal memory capacity will also have been distributed throughout the system. At one extreme, a stand-alone workstation – whether or not it forms part of a larger network – must have enough internal memory to handle the full range of CAD tasks. At the other, an intelligent terminal connected to a host computer will typically have sufficient memory to store the programming for routine graphics functions and the data needed to display one view of a model. The model itself, together with the rest of the system's programming, will be stored centrally.

In all except the very smallest microcomputers there will be a need for some form of back-up storage on which programs not in immediate use can be kept at the ready and on which data can be filed while remaining available for recall as and when required. Today, such back-up capacity is almost certain to be provided by discs of one sort or another. These range from hard discs with very large capacities, and very fast access times, to the familiar floppy discs which are used in all sorts of micro-based systems such as word processors. Access to all discs is measured in milliseconds (thousandths of a second), and this means that the user may well not be aware that the computer is having to 'look up' a program or a piece of data. Indeed, this speed of access has led to systems such as disc drives, which are essentially storage devices outside the computer (as opposed to memory inside it), being described as 'virtual memory' – the computer can refer to their contents so fast that it is almost as if it was in the memory proper.

But information in virtual memory devices must not only be 'found'; it must also be transferred into core memory and, when finished with, updated as necessary and put back on to the disc. It is clearly vital that the computer keeps track of what is stored where. This task together with the smooth organization of the traffic between memory and disc and memory and central processor, is a function of the operating system, the basic program that handles all the 'administrative chores' essential to the computer's capacity to run any software, whatever its nature or purpose.

The terminology of computer memory and/or storage is about to be further complicated by the arrival of yet another device, the so-called 'silicon disc'. Essentially this is a microchip very similar to those which provide the random access capacity of core memory; but the important point is that it is connected up to the machine, and handled by the operating system, just as if it were an orthodox disc drive unit. In the view of some people, the fact that the cost of integrated circuits looks set to continue its headlong fall for another decade or more suggests that silicon discs will eventually render the discs that are now being used obsolete and, in effect, all the information in a system will be treated as if it were in core storage. In the mean time, the advantage of silicon discs, other than their likely fall in price, lies in the speed with which they can be accessed. Although a gain of a few milliseconds might seem neither here nor there, it can make all the difference between a system which appears eager and responsive and one which has designers tapping their fingers with impatience.

Currently, however, most microcomputers rely upon floppy discs for virtual memory capacity while minis rely upon hard discs. There are also, at the bottom end of the scale, various 'diskettes' or 'mini-floppies', some usable only with particular models of microcomputer, and, in between, 'minidiscs' which are smaller versions of the standard hard discs used in a full-sized system. To some extent a number of different devices may be found within the same system: a terminal with its own built-in microcomputer, for instance, may also have a local storage facility in the shape of a floppy disc, while the minicomputer to which it is connected may have one or more hard disc drives. But, especially when a system is being built up over a period of time, it is essential to keep a wary eye open for potential incompatibilities between different forms of back-up storage.

Information stored on disc, whether it be programming or data, can of course be taken out of the system altogether, 'off-line' in computer jargon, and filed away for future reference. Thus discs containing special-purpose programming need not constantly occupy precious storage space in a disc drive unit, and data which is not currently being worked on can be removed to make way for others jobs. But discs have several drawbacks as an archival medium for storing information over the long term.

They are, first of all, expensive – in the case of hard discs significantly so. Unless they are stored under strictly controlled conditions the information they contain will be degraded or totally destroyed. Hard discs must be kept permanently in their airtight containers and are, therefore, extremely bulky. For this reason, many CAD manufacturers offer, as an option at least, the facility to transfer information on to magnetic tape for more permanent storage.

It is, of course, no more than 15 years since tape was the commonest form of back-up storage – remember how ranks of upright drive units with tape moving jerkily back and forth from reel to reel were almost a symbol of computing in the 1960s? But, in comparison with discs, tape offers access times which are painfully slow and it is now used only in cassette form on a few of the smaller microcomputer systems. However, as a means of recording information *en masse* cheaply and easily, and storing it comparatively economically and securely, tape has many advantages.

It must be said, however, that it also has some fairly major disadvantages. There is, first of all, the problem that making even minor additions or modifications to data will almost always involve recopying an entire tape, for information on a tape, unlike that on a disc, must be recorded strictly in sequence. In the context of CAD this can be a severe handicap. A busy drawing office may be constantly making minor modifications and additions to existing designs and appalling confusion can arise if there is not an adequate system for controlling the way in which these are recorded and incorporated in archive material.

There is also the fact that, in physical terms, tape is by no means as permanent a storage medium as it might at first appear. Even when wound tightly on a spool it will, over a period of years, stretch and sag. Although the distortions in the plastic tape itself are infinitesimal, they are quite sufficient to damage and ultimately destroy the tightly packed data recorded in the magnetic coating. The 'life' of tapes can be extended, it is true, by regularly turning them on their racks, rather as the *vignerons* of the champagne country regularly turn the bottles of maturing wine in their *caves*, but sooner or later, if the archive is to be preserved, it will be necessary to recopy the material onto new tapes.

There is, of course, one other way in which information can be stored – on paper. We will be discussing the whole question of 'hard copy' equipment at the end of this chapter; for the moment it will be sufficient to point out that, once transferred onto paper, graphical information is no longer accessible to a computer (using contemporary technology, at least) and can only be reinserted into the system with much painstaking work on the part of a human being!

Machine Communicates with Man: The Graphics Display

The display screen (see Figure 36), the window through which a designer peers into the electronic 'mind' of his computer-collaborator, looks very little different from an ordinary domestic TV set as far as the layman is concerned. And it is, of course, basically the same kind of mechanism: a cathode ray tube (CRT). But the differences, although not obvious, are in some cases highly significant.

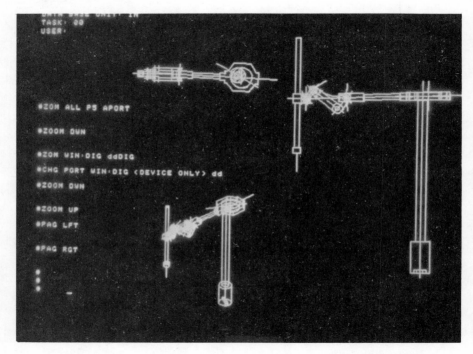

Figure 36. *A typical CAD monitor screen, showing the display of both graphics and text.*

All CRTs have three basic components: an electron gun which 'fires' a stream of charged particles at the back of the screen, a mechanism which allows the beam to be deflected or steered over the surface of the screen, and the screen itself which is coated with light-emitting substances called phosphors. When the electron beam strikes the back of the screen the phosphors at that point will glow.

A domestic TV employs what is called a 'raster display system'. The beam 'scans' the screen in a regular fashion, moving from the top left-hand corner to the bottom right, line by line, just as the eye scans a page of print, and the picture is made up of a series of dots or pixels. In the case of a domestic set, the image contains about a quarter of a million pixels (around 500 lines each made up of 500 dots). Changing the intensity of the beam as it moves from pixel to pixel allows changes in tone to be represented, and in a colour set different beams are used for each of the three primary colours and a system of masks built into the screen allows each to be focused only on the layer of phosphor which emits light of the corresponding colour. In order for the image to appear flicker-free to the viewer the beam of electrons must scan the screen at least 25 times a second.

It is, however, only comparatively recently that raster display has been adopted on a wide scale for CAD purposes. Most older systems, and some contemporary ones, use the elements of the CRT in a quite different way.

But before we consider the three main display techniques, it is worth considering what, from the user's point of view, are likely to be the most important characteristics of a display system. The main factors can be summarized as follows:

★ The display must be as flicker-free and place as little strain on the eyesight as possible. Concentrating hard on the screen of a CRT for hours on end is very different from watching a TV show. If detail is blurred or the image is not consistently bright, the consequences for the designer's health, not to speak of the quality of the work produced, can be severe.

★ Response must be fast – 'drawing' on the screen should be at least as easy as, and if possible, easier than, drawing on paper.

★ It should be made as easy as possible for the designer to 'see what he is doing'. In practice one of the best ways of achieving this is by incorporating a moving cursor in the display (see below, page 74).

★ The freedom to use tone and colour are clearly valuable, and in the case of solid modellers (see next chapter) virtually indispensable.

★ Finally, although this is strictly speaking something that is more directly relevant to the needs of the system than to those of the user, the routine task of maintaining a display should make the least possible demand upon the resources of the computer.

Stroke-writing Display Systems

If one simply wants to 'draw' on a CRT screen, then the most obvious technique to employ is not a raster display but what is known as stroke writing or vector graphics. In a stroke-writing system the electron beam is not deflected to follow a regular scanning pattern but is instead steered around the screen very much as a pencil is steered around a piece of paper –

with the difference, of course, that the mathematical model in the computer's memory allows it to draw 'freehand' with perfect precision.

The great advantage of stroke writing is that it allows the designer to 'draw' lines which the system will then display with total clarity and, in principle, infinite precision. This contrasts with raster systems, where lines will always have to be approximated by a fixed grid of pixels and, in the case of curves in particular, may appear blurred or ragged for that reason. But in a stroke-writing system, the only limit upon the complexity of the shapes that can be drawn and the accuracy with which lines can be displayed is the engineering of the beam deflection equipment, and in practice a screen image can achieve all the precision which the most perfectionist designer could require.

There are, however, disadvantages to stroke writing. It is not suitable for reproducing a wide range of tones, or indeed any image incorporating solid blocks of tone, and it is totally unable to provide full-colour images since the masks used in colour displays must be based on a rigid grid. Two further problems are peculiar to the original stroke-writing system – called direct beam refresh – though it also has some distinctive advantages which cause it to remain in quite widespread use.

As the name suggests, direct beam refresh uses the same beam of electrons to first 'draw' an image and then constantly 'refresh' it. In effect the beam is like a runner travelling round and round an intricate course, racing back to the starting point at the end of each lap in time to refresh the first part of the image before it begins to fade. This arrangement is highly satisfactory in that it allows for a very fast response. If the operator adds an entity, or deletes one, then the image on the screen will be amended just as soon as the computer has completed its calculations and without the other entities in the image being affected. It also means that a cursor can easily be displayed and moved about the screen and, in those cases where sufficient computer power is available, it allows an image to be convincingly 'animated' with the help of kinematic programming (see Chapter 5).

The difficulties that arise with direct beam refresh are twofold. First, since the beam's path must be kept constantly under computer control, it requires a disproportionate amount of computer time in comparison with other systems. Second, the speed at which the beam can travel is constant and is limited by the performance of the computer and the deflection mechanism. This means that, as the amount of information (the number of graphic entities and other material) on the screen increases, so the number of 'laps' of the display which can be completed each second falls. Once an image becomes so complex that the beam can no longer complete its journey and return to the start within a sixteenth of a second, the 'flicker' will soon increase to the point at which the display will no longer be of an acceptable quality.

One way of remedying both these problems is offered by a development of the basic direct beam display version of stroke writing known as direct view storage tube, or DVST. In a DVST system, the electron beam 'draws' the image once only. But the screen is coated with special persistent phosphors which, once activated, will continue to glow for as long as a second electron gun, the 'flood gun', bombards the entire screen. Clearly, this economizes on the system's demand for computer time and also allows it to cope with displays of unlimited size; once 'lit up' an image will continue to maintain its brightness indefinitely.

Unfortunately, DVST also has some rather severe disadvantages; these are mostly due to the fact that once an image has been traced on the screen no part of it can be deleted separately. A change which involves erasing some part of the image (which means any change where a line or any other entity is to be removed, changed in size or shape or simply moved) can only be achieved by clearing the entire screen and starting afresh with a new image. This is, obviously, a real handicap. Not only does it make it more difficult for the designer to retain the illusion of 'drawing', it makes it impossible for a screen cursor to be used. For slightly different reasons, it is also impossible to use a light pen (see page 73) with a DVST system.

Raster Display Systems

At first sight, since a raster display involves the computer in controlling the appearance of the entire screen area – each pixel must have a value assigned to it even if it is representing no more than a bit of the blank background to an image – it might seem even more demanding of computer time than either of the stroke-writing systems. But the fact that a raster display is divided up into a regular pattern of pixels, which remains constant no matter what image appears on it, makes it very much easier to handle computationally.

The essential point is that the computer does not communicate directly with the display control system; instead it calculates a 'bit map', a pixel-by-pixel description of the image to be displayed, which is then stored in a specialized block of RAM elements, the buffer memory, from which the display control draws its information.

In a sense, therefore, a raster system combines the best of both stroke-writing systems. Like the direct beam refresh system, it constantly re-draws the image, allowing changes to be incorporated rapidly whether or not they involve deletions. But like the DVST system it only requires the services of the computer when new instructions are issued, making it necessary to amend or replace the information in the bit map. Moreover, again like the DVST system, it ensures that the quality of the image will remain constant no matter how much information it contains. Raster technology has the further advantage of coping with full colour and of handling solid blocks of colour or tone just as easily as single lines.

The drawbacks essentially come down to a question of cost. If the display is to be of good quality the resolution will need to be high (that is, the number of lines and the number of pixels per line must be high) and the size of the buffer memory will have to be correspondingly great. To take a fairly typical example, a good quality (but not the best available) colour display will contain over a million pixels, each being allotted 24 bits of information to describe its colour and intensity. In order to store a bit map for that display, a buffer memory will need a capacity of around 3,000 kilobytes, roughly ten times the RAM capacity of a medium-range microcomputer!

In the early days of CAD this factor made it impractical even to contemplate the use of raster display. But silicon memory chips, which arrived in the early 1970s hard on the heels of the first microprocessor or logic chips, transformed the situation. Today raster display is indisputably the dominant technology, used in systems which range from the off-the-shelf micro with a display screen containing only a few thousand pixels (the Apple II screen, a fairly typical example, has a 279×159 pixel grid) to those on the frontiers of technology. Currently these frontiers are at 16 million pixels in the case of black and white and four million in the case of colour. These figures seem, for the moment, to be about the limit of what is possible. Masks for use in colour screens with higher resolutions would be so fine that there would be insufficient material between the holes to keep the thing together!

In general it can be said that the price of a raster system rises in proportion to the quality of its resolution. But although doubling the density of pixels per line results in twice as good an image, it must be remembered that doubling the number of pixels per line means roughly quadrupling the total number of pixels and hence the capacity of the buffer memory and the amount of processing required to 'repaint' the screen. So doubling the performance of a system increases its price by a factor which is nearer to four than to two.

It is worth bearing in mind, incidentally, that a system based on raster display will always 'know' more than it can show (it is likely to contain, in other words, a geometrical model that is a great deal more precise than the image on the screen which, even in the best system, will only approximate to what the model would actually 'look like'. You can of course always 'zoom in' and look at a more accurate 'close up' of an area of the image, but this may not always be an ideal arrangement.

How highly you value the quality of the display, and how much you are prepared to pay for it, will very largely depend on the applications which you have in mind. At the very basic level of graphics ability displayed by the average off-the-shelf micro, the low resolution can be a serious handicap. If a model becomes complicated then lines soon begin to overlap, and the image becomes unintelligible. All lines will appear to be to some extent 'dotted' and even in systems with much higher resolutions the 'step effect' in curved or diagonal lines can be quite marked (see Figure 37). 'Jaggies', as they are called, are indeed a perennial problem with raster displays and some systems employ an algorithm which attenuates the effect and makes it less obvious, but only a very high resolution raster display will produce curves and diagonals which appear as 'clean' as those generated by stroke writing.

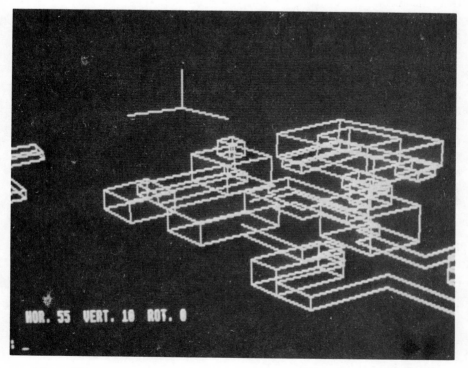

Figure 37. *Because the pixels which make up the image in a raster display are arranged in a regular grid pattern it is inevitable that the appearance of curved or diagonal lines will suffer from the step effect, or, more familiarly, be seen as 'jaggies'. Although there are algorithms which smooth out the worst jaggies, only systems with very high resolution will be totally free of the problem.*

At the other end of the scale there are few applications for which the really high resolution displays (those, incidentally, which the CAD manufacturers often like to use when they produce their showpiece demonstrations) would be really necessary – unless, that is, you are planning to venture into the field of computer-generated special effects or computer art. In between the two extremes, in the range between the 500 or 600 pixel-per-line standard of a domestic TV set and the 1000 pixel-per-line level at which prices really begin to soar, there are many off-the-shelf display systems available which will probably be more than adequate for most needs.

Man Communicates with Machine: Menus and Input Arrangements

The display system is of course the principal means by which the computer communicates with its human masters – in effect it allows the machine to put its geometrical 'thoughts' on view. But it is also necessary for the designer to communicate with the machine and, since we cannot make our thoughts visible and the computer would not be capable of understanding them if we put them into everyday language, it has required a good deal of ingenuity on the part of those who design and build CAD systems to make such communication both easy and effective.

In theory, of course, a designer could tell a graphics computer everything necessary via an orthodox keyboard. But this would involve the use of a high-level language such as FORTRAN or Pascal and hence much specialized training and a great deal of thought on the designer's part every time he wanted to issue a new instruction. In practice, therefore, the keyboard may be used comparatively little and then usually in combination with the two further channels of communication which, in one form or another, are incorporated in every CAD system. There is, first, a channel for conveying the designer's responses to a menu (virtually every CAD system is menu driven in routine applications) and, second, a channel which allows the designer to 'point to' things (particular points or areas, entities or parts of entities) on the display screen. The way in which these three channels, the standard alpha-numeric keyboard, the 'menu response' and the 'pointer', are used may become clearer if we show some simple examples where each part of the message is conveyed by a different channel.

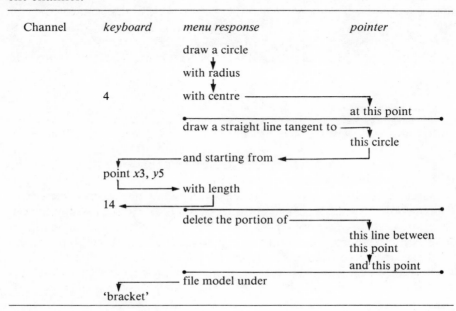

Channel	*keyboard*	*menu response*	*pointer*
		draw a circle	
		with radius	
4		with centre ——————————→	
			at this point
		draw a straight line tangent to ——→	
			this circle
	point x3, y5	←———— and starting from ←——	
		→ with length	
14 ←—————			
		delete the portion of ——————→	
			this line between this point
			and this point
	'bracket'	——— file model under	

Although there is an almost bewildering variety of different 'pointing' devices and several methods of presenting the menu and responding to it, all systems fall into one of three categories:

- systems where the input is via a light pen
- systems using a screen cursor which is moved by steering it around with one of a number of different input devices
- systems which make use of a graphics tablet

Within each of these three general groups there will still be wide variations. In some cases, for example, a light pen is used both to point to a position on the screen and to indicate the response to a menu (which may itself be displayed on the same screen as the graphics image or on a second screen), but in other cases the light pen will be used only as a pointer and the responses to the menu may be signalled via the keyboard or a smaller keypad, and so on; the range of different combinations is almost infinite – or so it sometimes seems.

Light Pen Input

Superficially at least, the light pen might seem the easiest system. For the pen, a small electronic device on the end of a cable which can (when pointed at the screen as shown in Figure 38) detect the scanning electron beam and thus signal its location to the computer, is easy to use and the analogy between drawing on a piece of paper with a pencil and drawing on the screen with a light pen seems satisfyingly literal. In really sophisticated systems quite surreal feats can be performed at the touch of the pen. Given a realistic representation of a motor car, for example, it might be possible actually to 'pick up' a front wheel with the light pen and place it to one side, or to ring some detail on the vehicle and find that the system instantly zooms in on it.

But there are drawbacks. Some people do not find that the use of the light pen comes naturally to them. It can be difficult, wielding what tends to be a rather clumsy implement, to point with any precision. This means that though it is indeed possible, in some cases, to 'draw' on the screen it is seldom a very practical technique. If the pen is simply being used to identify entities – 'the end of the line here', 'the centre of this circle', etc – the imprecision is less important, for most systems are able to work out that you mean this nearer line rather than that more distant one even if you are in fact pointing at the space between the two. It must also be said that leaning forward to point at the screen is perhaps a less convenient and more tiring business than the sort of remote control which characterizes the other two input techniques.

As has already been mentioned, some systems which use a light pen as a pointer may also rely upon it as a means of signalling the designer's

Figure 38. *Using a light pen a designer can point at a position on the screen, or at one of the entities already being displayed, in much the same way as a pencil might be used to point at a drawing. The light pen may also, in some cases, be used to select options from the menu which is sometimes incorporated in the graphics display or may, as in this case, be presented on a separate alphanumeric screen.*

responses to a menu. But in other instances the keyboard or a keypad (in effect, a small keyboard with perhaps only nine buttons) may be used to signal which option is being selected. A typical sequence of menus for general-purpose graphics work is shown in Figure 39. In either case the menu itself may be incorporated in the graphics display or presented on a separate screen – often a smaller monotone display limited to the presentation of alphanumeric data.

Finally, it must be remembered that a light pen is useless in a system which employs a storage tube display for, once the image has been 'drawn', there is no scanning electron beam for the pen to detect.

Cursor Steering Input Devices

A cursor, which can be steered around the display by a variety of means, offers an alternative to the light pen. It may initially seem to provide a less natural means of pointing at an entity or indicating a spot on the screen,

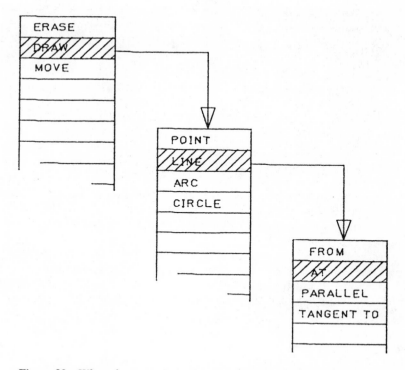

Figure 39. *When the menu is displayed on the screen, rather than being incorporated into a graphics pad, conveying a complete instruction will normally involve the designer responding to a sequence of menus; this example shows the first three options needed to convey the start of an instruction: 'Draw Line at . . . '*

but the necessary skill is quickly mastered and there is the advantage that the operator can sit back with the input device positioned wherever he or she finds most comfortable and convenient. The cursor is also, if anything, a more accurate and reliable input device than the light pen.

If we ignore, for the moment, those systems which use cursors in conjunction with a graphics tablet (see below) there is a choice of three guidance devices currently available: joysticks, tracker balls and the 'mouse'. In each case, as well as the actual 'steering' system which moves the cursor around the screen there will also be some means of 'tagging' a particular spot, rather as one might place a dot on the paper with the tip of a pencil. This not only allows the operator to manoeuvre the cursor into precisely the right position before signalling to the computer that *this* is the desired position for, say, the centre of an arc or that *that* is the line concerned; it also makes possible techniques such as 'rubber banding' where one end of a line is 'fixed' in position and the other remains attached to the cursor so that the line itself swivels and stretches like a rubber band until the cursor has been satisfactorily positioned and the other end of the line can be tagged. In some cases the input device itself will incorporate a tag button; in others the necessary signal is delivered via the keyboard or keypad.

Anyone who has played computer games will be familiar with the use of a joystick device. Its main advantage over the other two devices is that it is comparatively easy to control not just the direction in which the cursor moves but also its speed. Thus, large movements from one side of the screen to the other, for example, can be completed rapidly while more delicate manipulation of the joystick will still allow minor adjustments to be made with a high degree of precision. The joystick is, on the other hand, self-centering so that the operator's hand must be kept on it permanently if the cursor is to be kept in any particular position. Some systems avoid this problem by allowing the position-sensitive mode to be switched to a rate-sensitive one in which the operator can control the speed with which the cursor moves.

The tracker ball is, as the name suggests, a ball mounted so that its top half can be rotated freely in any direction. It is, in fact, not unlike a joystick unit from which the stick has been removed. Since the cursor can be 'parked' anywhere on the screen by simply leaving the ball in position, it allows the cursor to be manoeuvred with great accuracy. But it is also true to say that large-scale movements are achieved more slowly and less 'naturally' than is the case with the joystick.

Although the introduction of the mouse as an input mechanism for general-purpose computer applications has been trumpeted as marking something of a revolution, it cannot really be said that it offers any great advantages over the other two devices where CAD is concerned. In fact, the mouse is probably best thought of as a tracker ball turned upside down so that the ball is turned by moving it this way and that over a flat surface. In principle this combines the speed of the joystick with the precision of the tracker ball, but the need to keep the mouse orientated consistently so that its up and down and left and right are the same as those on the screen can be tiresome, and many people find that the whole arrangement is something of an acquired taste.

Whichever of the three devices is employed, it will of course be used in conjunction with a menu and a keyboard or keypad. The cursor can also be used to select responses to the menu if it appears on the graphics screen – the more usual arrangement – but if the menu appears on a second screen the operator's choice must normally be signalled via the keyboard or keypad, though some systems allow the cursor to be moved back and forth between screens.

Which of the three devices is selected will remain largely a matter of personal preference, and one of the easiest ways in which manufacturers could make their hardware more widely acceptable would be to so design it that any one of the three could be used rather than, as is at present the case, presenting the buyer with one of them as a *fait accompli*.

Graphics Tablet Input

The third of the three main input systems differs from the other two in that it involves the use of a graphics tablet, a 'desk-top' replica of the display space which includes both a blank 'screen area' in the centre and a menu space which adjoins it (see Figure 40). Using a 'pointer' device the designer can draw on the graphics tablet or indicate entities or positions on the screen (which incorporates a cursor that 'follows' the movements of the pointer as it travels around the tablet) as well as 'picking out' instructions from the menu area of the tablet.

Figure 40. *A typical CAD terminal incorporating a graphics tablet for input. By moving the pointer or stylus around the blank area in the centre of the pad in front of him the designer can position a cursor on the screen, and by pointing at the surrounding menu area he can convey instructions such as 'draw', 'erase', etc to the system. Other instructions or data such as dimensions, etc will be input via the keyboard or a smaller keypad.*

The tablet itself is a flat surface with a grid of wires beneath it which picks up signals from the pointer whenever the operator presses a tag button on it. The pointer may be one of two varieties: a pencil-like stylus or a digitizer. The latter, sometimes called a 'puck', looks rather like a small magnifying glass incorporating a crossed hair 'gunsight'. Using a digitizer it is possible to transfer graphical information from a drawing on paper into computer memory. The graphics tablet will therefore be almost the automatic choice for any user who anticipates the need to convert data from one medium to the other on a regular basis.

It should also be added that menus can be changed relatively easily; each set of applications software for a system will have its own menu and switching from one to the other simply means that one sheet of paper, carrying one menu, must be removed from the pad and replaced by another whenever the computer program is changed. Figure 41 shows sections of some typical menus.

In comparison with light pen and cursor steering systems the graphics tablet has several quite important advantages. It requires neither the use of a second alphanumeric screen nor the sacrifice of precious space of the display screen in order to present the menu. The menu itself can be used rather more economically since the designer can issue a set of instructions by picking them out one after the other rather than having to work through a hierarchical sequence of choices. It is, in terms of the hardware and software required, far and away the most economical system and is in practice the usual choice that is available when a general-purpose microcomputer is being employed. For, unlike the light pen, which requires specialized hardware in the display system, and cursor steering devices, which demand considerable processing power if they are to operate in real time, graphics tablet operation can be achieved with standard display equipment and relatively simple software.

Provision must, of course, be made for input of other instructions or data and, as with the other two systems, this will normally be achieved via either the main keyboard, or in some cases a specialized keypad that allows the rapid signalling of numerical data may be built into the digitizer itself.

Choosing an Input System

There is much controversy, even within the CAD 'community', as to the pros and cons of different input devices, different methods of displaying a menu, different ways of inputting alphanumerical data and instructions – in short, all the various means and combinations of means by which man and machine can communicate.

In a sense, much of the argument is likely to be of little interest to the purchaser of a turnkey system, which means the vast majority of CAD users. For most manufacturers have opted for one particular set of input arrangements and, all other things being equal, it is unlikely that this feature will be the one which determines the choice of a system. The fact that manufacturer A provides a system with light pen input while manufacturer B provides graphic tablets, for example, may superficially seem to be a very significant distinction between the two, but in practice it is

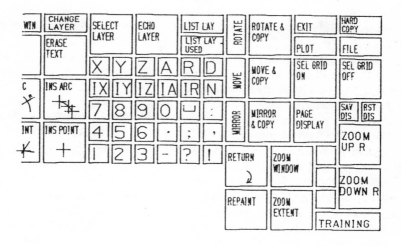

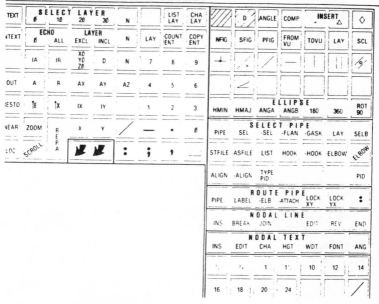

Figure 41. *A section from the menu area of a graphics tablet – the tablet will also incorporate a blank area corresponding to the display screen itself. The top example is taken from a typical general-purpose graphics menu while the bottom one shows a section of the menu designed to be used with special-purpose software employed in designing pipework. When one computer program is exchanged for another, it is necessary only to take one menu sheet off the graphics tablet and replace it with another.*

likely to be comparatively unimportant compared with, say, the range of software available, the average speed of response or the amount of memory capacity.

With the exception of those few cases where a particular input arrangement has unique advantages (as is the case, for example, with the graphics tablet-digitizer combination when dealing with the transfer of graphics information from paper to computer) or clear limitations (as with the impossibility of using light pens with storage tube display), the choice is largely a matter of personal preference and, of course, economics.

There are some manufacturers, for instance, who argue fiercely in favour of twin display screens. Even if one is used solely for alphanumerical display for 90 per cent of the time, they suggest that the fact that it can also be used for a second graphic display makes the system as a whole vastly superior to those which rely upon graphics tablets or smaller monotone display screens to present the menu. This is indeed the case, but it is also true to say that such display systems are a good deal more expensive and the additional graphics display may, in most applications, be a luxury that is useful but not indispensable. Those manufacturers who have opted for single-screen systems argue that they cause less fatigue since the designer's attention is not constantly switching back and forth from one screen to the other.

From the users' point of view it is ultimately not so much a question of trying to choose the input combination that is ideal as of trying to make sure that you avoid any which have serious drawbacks in your own particular circumstances.

Plotters and Other Hard Copy Devices

For the foreseeable future at least, until some much wider and more universal channel of communication is opened up between CAD and CAM, and until display screens are widely distributed around the average factory floor, every user of a CAD system will want the option of converting information created and stored inside a CAD system into a set of plans on paper. As an archival storage medium, a means of communicating graphical information over a distance and recording it in a form which allows it to be studied at leisure, the traditional plan or blueprint still has much to commend it. It can be popped in an envelope and posted around the world, or pocketed and carried round on site; it can be duplicated quickly, easily and cheaply; it can be examined without the need for any special equipment by one's fireside at the weekend, say, or half-way up a mountain in the wilderness.

The choice of a plotter or hard copy device (the confusing use of 'hard copy' to describe a piece of paper is presumably justified by contrast to the even 'softer' copy which exists inside the computer) is both an important issue in its own right and significant in that it may affect the overall performance of the system and the convenience with which it can be used.

Until comparatively recently all plotters could be classified on two fundamental bases, and although the introduction of exotic technologies such as lasers have complicated matters somewhat this still holds good for most practical purposes. There is, first of all, the question of how the lines are actually transferred on to paper and, secondly, the question of the kind of paper that is used.

Figure 42. *A pen plotter.*
(Photograph reproduced by courtesy of Computervision)

The most accurate and elaborate plotters, which are also normally the most expensive, 'write' with a pen – or pens, if more than one colour or weight of line is required. The great advantage of pen plotters is that they can achieve almost unlimited precision and a very high degree of legibility; their principal disadvantage is that they are comparatively slow. The only alternative in widespread use is the electrostatic plotter, which is very much faster but often less accurate and nearly always incapable of matching the clarity and legibility of a pen plotter.

Electrostatic plotters operate on a raster system, building the drawing up out of a series of dots – their high speed is due to the fact that the drawing is constructed by scanning the paper exactly as the electron gun scans a display screen, line by line from top to bottom, rather than drawing each entity in turn as with a pen plotter. But their use is not limited to raster display systems since they generally take their information direct from the computer database and not from the buffer memory in the display unit. This means that even with a raster display, the plotter may work to a far higher resolution than the screen. An exception to this rule are the desk-top hard copy units which are used to produce a quick copy of the screen image, the equivalent, if you like, of a xerox, which is relatively crude but convenient for many purposes. In some cases, these units are thermal rather than electrostatic copiers, in which case they require specially prepared paper.

The handicap imposed by the slow speed of pen plotters can be overcome to some extent by equipping them with their own microcomputer control units, thus transforming them into 'intelligent plotters'; this means that the graphics computer can 'unload' the geometrical model into the plotter's memory, thus freeing the computer of the chore of driving the plotter and allowing the resources of the system to be used for other purposes, a significant gain but one that is of limited advantage if what is wanted is a drawing that will be quickly available.

The second basis for categorizing a plotter depends upon the type of paper in use. Almost all electrostatic and thermal machines are fed from a continuous roll of stationery which moves past the row of 'writing heads' each of which covers one small fraction of its width. Pen plotters, however, may be either flat-bed machines using individual sheets of paper or drum-fed devices using continuous stationery. The very best results, undoubtedly, are obtained from flat-bed pen plotters which also have the merit that the 'bed' can, in principle, be of almost unlimited size whereas the width, though not the length, of the plans produced by a belt-fed machine will always be comparatively restricted.

But flat-bed plotters, which usually have the further advantage of being able to utilize standard stationery, have at least one severe handicap – someone has to be there to remove one sheet and replace it with another every time a task is completed. Drum-fed machines, in contrast, can continue to work unattended for as long as their paper supply lasts. In practice, this means that a system with a drum-fed plotter can often be left to 'print out' hard copies overnight when the system is otherwise unutilized, whereas a flat-bed machine may demand the attention of both the graphics computer and staff when both could be usefully employed elsewhere. The distinctions, however, are not always rigid. Some drum systems use cut sheets of paper and some flat-bed machines may use continuous stationery.

Several new developments that are likely to widen the range of choice of plotters are already far advanced. Laser printing, currently in use for specialist applications such as cartography, looks likely to become cheaper and more generally available; it combines the high precision of the pen plotter with the speed of the electrostatic system, but is still expensive and unable to cope with a wide range of colour. In the US, and increasingly elsewhere, another possibility is computer on microfilm (COM), a system which uses fine beams of laser light to draw directly on to standard 35mm film. The speed with which these machines work and the high degree of precision achieved, coupled with the convenience of the final product, a transparency which can be stored, duplicated or transported as easily as a holiday snapshot, suggests that they may well become extremely popular (see page 138 for further discussion of these developments).

Chapter 4
Principal Types of CAD System

As we saw in the last chapter, the range of equipment used in CAD is very wide and there are many different ways in which systems might be classified on the basis of their hardware components. But, from the user's point of view, the type of display technology chosen, the method of input adopted, even the way in which computer resources are deployed, are secondary considerations. The most important and fundamental characteristic of a system is the nature of the geometrical models which it is able to construct, which is, of course, primarily a function of the software rather than the hardware.

On the basis of their modelling abilities, all CAD systems can be assigned to one of four main categories: 2-D modellers, 3-D wire-frame modellers, surface modellers and solid modellers. There are, it is true, areas in which the divisions become blurred and the categories are not always mutually exclusive – the first three, indeed, form a clear progression in which a system in a higher category will normally share the abilities of a system in a lower category. But overall the distinctions are sufficiently clear-cut to form the basis of the discussion which follows. Even more significantly, as far as the user is concerned, they have a very direct bearing upon the performance of a system and the applications for which it will be suitable.

Two-dimensional Modellers

Any drawing can, of course, be described as a 2-D model of the object it depicts, and when dealing with the sort of graphical information used in engineering or architecture, say, it is not unreasonable to look upon a 2-D modeller as a sort of computerized drawing board. Many of the advantages which this most basic form of CAD offers over the more traditional apparatus in the drawing office may, therefore, be seen more as improvements in the degree of speed, efficiency and convenience with which tasks can be performed rather than a change in the nature of the design process itself. None the less, as we tried to suggest in Chapter 1, such improvements can be very significant in themselves.

The use of macros (see page 108), for example, or the ability to 'zoom in' and work on one detail of a model and then 'stand back' and consider it as a whole, can revolutionize a draftsman's work and may cut the time taken to deal with some design tasks by up to 75 per cent or more. Equally important is the freedom which the use of CAD gives for different 2-D models to be combined or juxtaposed. In effect, the system will allow the use of a virtually unlimited number of overlays which can be superimposed, moved about and removed again in any order and without any of the fiddle and lack of clarity which are involved when sheets of paper are laid over each other.

There are, indeed, many applications in which the capacity to deal with a third dimension would be a wasteful and even a counterproductive elaboration. The design of printed circuit boards, for example, or even integrated circuits demands no more than the ability to superimpose the lay-out of the two sides of the board or successive layers in the chip in order to establish that connections are accurately aligned. Similarly, those using CAD methods for cartography or surveying work may find the close parallels between a 2-D system and more traditional methods to be a positive advantage, allowing data to be transferred from paper to computer memory and vice versa with complete freedom.

But with a little ingenuity it is perfectly possible for some 2-D systems to be developed into what are sometimes called '2½-D' modellers. For, although a series of 2-D models contain no explicit information about the third dimension, they may, if they represent a succession of regularly spaced slices through a single object, contain a good deal of implicit information. In effect, each 'layer' in the model is equivalent to a contour line on a map and by adjusting the intervals between different layers it is possible to use a 2½-D modeller to specify a 3-D object with varying degrees of precision.

It may even be possible to transform all the layers in order to obtain what is, in effect, a perspective view of the entire model. Figure 43 shows an image of this kind. Again, there are many applications in which a system with this sort of ability is perfectly adequate for all needs. Many machining operations, for instance, only involve specifying the tool path in two dimensions together with the depth to which the tool is required to plunge. By carefully selecting the appropriate separation between layers of the model and by a judicious choice of curved or bevelled cutting tools it would even be possible to use a 2½-D modeller to create the tool path required to machine the smoothly contoured pocket shown in Figure 44.

The advantages of 2-D modellers, and their 2½-D relations, are simply stated: they make far lower computational demands, in terms of both processing power and memory capacity, than any other type of system; personnel already skilled in traditional design techniques find it relatively easy to adjust to their use; by automating many of the more tedious and repetitive aspects of the design process, including the preparation of NC

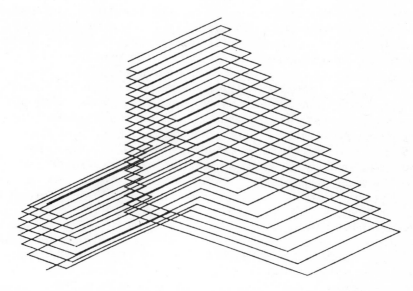

Figure 43. *A '2½-D model' of a wedge-shaped object with a cylindrical projection; this was generated by taking a series of 2-D models, each representing a 'slice' through the object, and 'layering' them on top of each other.*

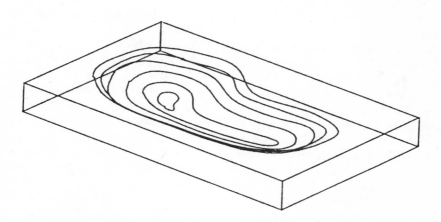

Figure 44. *The surfaces of the kidney-shaped pocket in this block are curved along all three axes. None the less, by choosing an appropriate distance to separate the layers of a 2½-D model, and by using bevelled cutting tools, it would be possible to create a NC part program to produce this object using a CAD system confined to 2-D modelling.*

programming, they can achieve dramatic improvements in efficiency and productivity – *in certain applications*.

But there are many other applications, probably a large majority in fact, in which the introduction of 2-D modellers carries with it the dangers that always accompany a revolution that is begun but left uncompleted. Essentially, the point is that if a designer's work (or any significant part of it) requires that he think in three dimensions, then it will ultimately prove a false economy to invest in a CAD system that is limited to 'thinking' in two. For, as we tried to show in Chapter 1, the benefits of having a mechanical collaborator which can deal with 3-D descriptions with the same ease as people cope with 3-D concepts are so great that it would be folly to miss any opportunity of capitalizing on them.

Wire-frame Modellers

A 3-D wire-frame model, like a 2-D one, is made up of lines linking specified points in space – with the difference that the space now has three dimensions. The fact that wire-frame modellers can be seen as extending traditional graphical techniques into a third dimension is, no doubt, one reason for their popularity. They also have the advantage that they can be implemented on any of the three types of display system and with any of the input devices. A further point in their favour is the fact that, compared with surface or solid modellers, they incur a relatively low computational overhead.

In essence, a wire frame is simply a drawing in which the position of the entities is specified in three dimensions rather than two. The basic principle (and, as we shall see, some of its drawbacks) can be illustrated by the simple example of the old-fashioned lampshade which often consisted of parchment or fabric attached to what was, literally, a wire frame. Figure 45 shows the sort of thing we are talking about.

A 3-D model, even when seen only on a 2-D display screen, has many advantages over a 2-D representation of the same object. To start with, it removes many ambiguities. For instance, a familiar optical illusion, the Necker cube which can equally easily be 'seen' in either of two ways (Figure 46), will cease to be ambiguous if drawn as a wire frame because the computer will always be able to transform it to present a view, a perspective perhaps, in which the ambiguities are resolved. The computer will remember, even if the designer does not, which of the sides is the 'front' and which the 'back', or, to put it more accurately, how the model is aligned in the computer's 'model space', the 3-D equivalent of the sheet of paper on the designer's drawing board.

A wire-frame modeller also allows information about all three dimensions to be incorporated in a single model and will have the capacity to transform that model in order to show what it 'looks like' from different

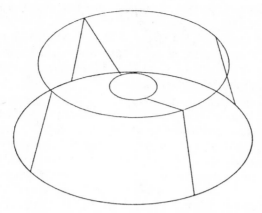

Figure 45. *A wire frame, such as might provide the rigid 'skeleton' for a lampshade made of parchment or fabric. This example illustrates the point that, although a wire frame is in some sense a drawing in three dimensions, it has severe limitations. If the model is rotated, as here, then the 'wires' will cease to correspond with the 'horizon lines' that delineate the edges of curved surfaces.*

points of view. At the minimum, the system will be able to display the basic orthogonal views along the *x*, *y* and *z* axes, an isometric view and, perhaps, obliques or perspectives. At the other end of the scale are systems which can present a model from virtually any angle or in any form

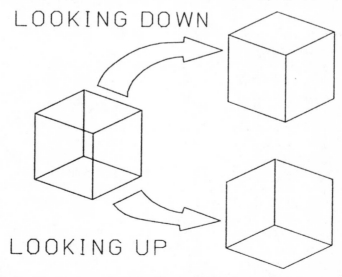

Figure 46. *The Necker cube, a familiar example of an optical illusion. When drawn on paper the cube can equally easily be interpreted in either of two ways, but drawn as a three-dimensional wire frame model it will be totally unambiguous as far as the computer is concerned.*

the operator cares to specify and which allow the designer to 'fly around' his creation as if in a helicopter simply by turning a knob.

The more elaborate the transformations (and the more complex the models they are applied to and the greater the speed with which they are effected) the greater are the computational resources and complexity of the software required to handle them. It must be remembered that each time a new image appears on the screen, even if images follow one another so rapidly as to simulate movement, the computer is taking all the basic geometrical data and using it to construct a completely new 'drawing'. When a model contains many hundreds of entities the amount of calculation involved in even quite a simple transformation is awe inspiring to human beings who lack the appetite even to nibble at the numbers which computers crunch in a mere fraction of a second.

But there are cases where viewing a wire-frame model in isometric or perspective, instead of resolving ambiguities and removing confusion, only serves to multiply them. If, for example, we took the object in Figure 43 and drew it as a wire frame we would produce something like the image on the left of Figure 47. It is by no means obvious that this represents the object on the right. For rather than drawing horizon lines to indicate the edges of the cylinder as they would appear from this particular viewpoint, the system displays one line linking the points of origin of the two circles and another joining their centres. Without the image on the right to clarify matters it might be far from clear what was being represented.

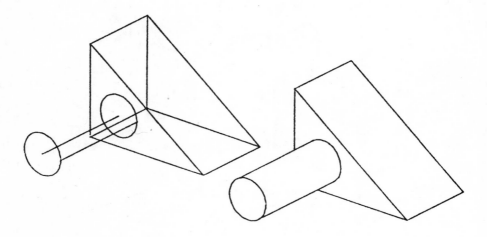

Figure 47. *The wire-frame model (left) may not provide a clear or unambiguous representation of the object on the right.*

There are, it is true, programs that now take care of this sort of problem when regular geometrical forms such as a cylinder are involved, but when dealing with objects that are made up of complex curves only a surface or a solid modeller can calculate the position of every point on the surface of those objects, their interrelationships and, therefore, where the horizon will appear under all circumstances.

Similar sorts of difficulties, arising out of the system's ignorance of the surfaces which link the entities in its wire-frame model, are encountered when elements of a model intersect with another or when it would be useful to construct a cross-section. Coming back to the cylindrical section of our model, for example, it is easy to imagine circumstances in which it might be important to establish the exact form and size of the ellipse that would be formed if we were to cut across the model along a particular plane, as in Figure 48. A wire-frame modeller, however, cannot help us here, for it knows nothing of the surfaces which would form the edges of the ellipse.

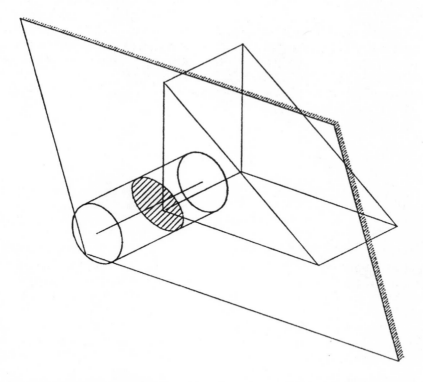

Figure 48. *If the cylindrical part of the model is cut along an oblique plane, a wire-frame modeller will not be able to tell the designer anything about the shape or area of the ellipse that has been created; it will, in fact, simply show the cut ends of the two wires and nothing further.*

Another problem with wire-frame models is that the amount of information they contain is as likely to be a source of confusion as of clarification.
Consider, for example, the wire-frame model of a relatively simple
object – the pulley block shown in Figure 49. Each line represents an
edge, yet with so many lines criss-crossing each other the observer is be-
mused and quite unable to perceive the surfaces that are being outlined.
The obvious remedy is to remove the hidden lines which would, in reality,
be invisible. The difficulties, however, are daunting, for in order to establish which lines, or parts of lines, will be visible and which invisible the
system must again determine the relative position of the surfaces and the
whereabouts of the horizon lines.

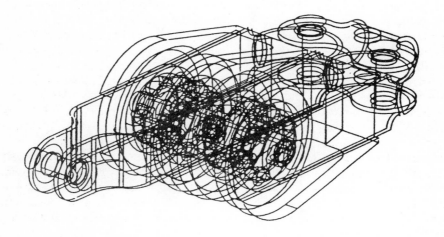

Figure 49. *A wire-frame model of even a relatively simple object, such as this
pulley block, can be profoundly confusing unless there is some means of
eliminating the 'hidden lines'.*

In some systems, the removal of hidden lines is an operation that can
only be performed manually, with the operator identifying the entities
which are to be deleted. This would, of course, not be an ideal remedy
even if it did not involve much painstaking labour, for as soon as the
model is transformed lines which were previously hidden will become visible and vice versa. A rather more sophisticated solution involves the use
of programs which work 'back' from the surface of the screen assuming
that, unless the operator has declared otherwise, every shape enclosed by
entities must represent a surface and eliminating everything that lies
'behind' them. There are also programs which work in the reverse direction, drawing the entities that lie furthest away first and then deleting
them if and when a surface is subsequently drawn in front of them.

These are time-consuming and, in terms of computer resources, expensive procedures. Another possibility is to vary the intensity of lines, making those that are 'near' the surface of the screen brighter than those which lie further 'behind' it. This is a very effective method of providing visual cues that help to clarify images that would otherwise be totally bewildering. Unfortunately, though, neither of the stroke-writing systems can generate lines of varying or graduated intensity and it is therefore only feasible with raster displays.

Three-dimensional wire-frame modellers obviously represent a considerable advance over the simpler 2-D systems. But at the same time the sort of graphics they produce are recognizably a development of rather than a departure from the traditional lines-on-paper descriptions which designers are accustomed to. Indeed, once the novelty of the gadgetry has worn off and the new-found freedom to draw in all three dimensions has been assimilated, the creation of a 3-D wire-frame model involves much the same techniques, on the designer's part, as the production of a set of drawings — with the difference, of course, that the computer takes over a good deal of the donkey work.

In many people's eyes the fact that wire-frame and modellers automate the drafting process and extend it into the third dimension without altering its fundamental nature is a distinct advantage — a switch to CAD involves the designer learning some new techniques, but not relearning his craft from scratch as is the case with some solid modelling systems. It is therefore important to realize the limitations of the wire frame and to understand that while it may be more than adequate for, say, an architectural designer, for whom orthodox line drawings may still be the best means of graphic communication, it will be of very little help to an aerodynamic designer, who is likely to be dealing with complex surfaces, or a designer of mouldings, who is likely to be concerned with the properties of solids.

It must always be remembered that although a human being almost automatically 'translates' a wire-frame representation into a solid object in his mind's eye, the computer does no such thing. It has absolutely no knowledge or understanding of the surfaces that span the frame or of the space which it encloses. This means, for example, that a wire-frame modeller can seldom tell us anything about what happens when two wire frames intersect. Figure 50 shows a very typical case in point, a wire-frame cube with a wire-frame cylinder passing through it obliquely — the sort of combination that is likely to be routinely encountered in, say, the design of a process plant. What the designer is likely to want to determine is, of course, the precise form of the surfaces in the areas where the two shapes intersect. But the wire-frame modeller cannot define the position (in terms of x, y and z coordinates) of any point on either of the two surfaces; it knows only about the position of the edges represented by its frame. It cannot even tell us, for instance, which bit of the cylinder lies inside the cube and which bits lie outside, for it does not know where the surfaces of the cube are.

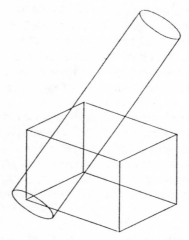

Figure 50. *A wire-frame cube with a wire-frame cylinder passing through it. A CAD system relying on wire frames alone can tell us nothing about the way in which the two objects interact.*

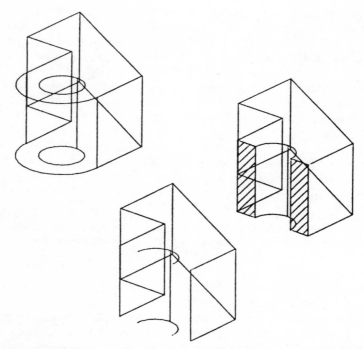

Figure 51. *The fact that a wire-frame modeller cannot produce the cross-section at the top right from the isometric view on the left graphically illustrates its limitations. If the model is cut in this way, all that will be produced is the 'nonsense object' at the bottom, with its projecting wire ends suspended in space.*

This limitation also makes it difficult for wire-frame modellers to handle cross-sectioning. To even a lay person it might seem child's play, given the isometric view on the left of Figure 51, to produce the cross-section at the top right. But from the computer's point of view, all that is left if its wire frame is cut along this plane are the ends of eight 'wires' suspended in thin air, as it were, as in the view on the bottom right of the figure. In this instance, of course, the designer can obtain the desired cross-section by joining up the ends, using his own understanding of the shape of the object being modelled. But what happens if we want a cross-section of a more complex object or specify a different cutting plane, as in Figure 52? Here, clearly, joining up the ends of the wires would require the designer to have exactly the information (the shape of the ellipses formed by the cutting plane) which he wants to obtain!

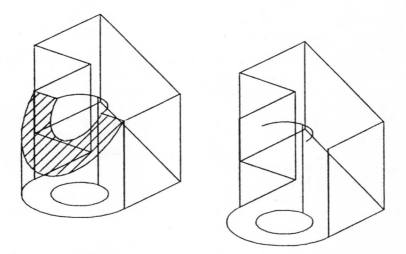

Figure 52. *Although it would be possible for a designer to make sense of the cross-section shown in Figure 51 by joining up the ends of the wires, no such remedy is possible when curved surfaces are dissected by an oblique cutting plane – though the shapes thus created will probably be precisely the information that the designer wants the computer to calculate.*

Surface Modellers

The obvious way of overcoming the limitations of a wire-frame model is to clothe the bare bones of the frame by laying surfaces over them, and surface modellers are indeed the next stage in the CAD hierarchy. Adding surfaces clarifies many ambiguities and makes explicit much information that with a wire-frame model could only be implicitly assumed by the designer. Figure 53, for example, shows a wire frame which, in its original form, is confusing to even an informed eye, and the same model clarified by the addition of surface meshes.

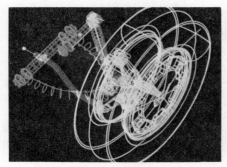 

Figure 53. *The addition of surfaces to clad the wire frame will often resolve ambiguities and clarify confusion; in this case the surfaces have been 'clad' in solid tones rather than a mesh and thus have an extremely 'realistic' appearance. (Photographs by courtesy of Computervision.)*

There are, clearly, many applications in which the ability to model surfaces as well as edges is not just helpful but absolutely essential. A wire frame is, for example, virtually useless as a means of representing the sculpted form of a car body and cannot possibly be used to provide the sort of description of it that would be needed at the manufacturing stage. Nor, to take an example from a totally different field, will a wire-frame modeller be of any use to a cartographer who wants to model the topography of a landscape.

There are also many cases where a surface-modelling capability is vital for the preparation of NC part programs – essential, that is, if the task is to be handled automatically by the computer system and is not to involve the programmer in creating what would virtually be a surface mesh by constructing a very elaborate wire frame.

Finally, most of the analytical techniques based on finite element analysis and boundary element analysis can only be employed with surface modellers, and some, of course, only with solid modellers, for dividing a model up into tiny elements (as shown in Figure 54) in order to examine the stress placed on each or the degree to which it will distort under that stress requires a precise knowledge of its surfaces and the underlying material.

Unfortunately, though, the mathematics needed to deal with a surface model are far more complex and, even more importantly, more time consuming than those which are used to construct a wire frame. To illustrate the point, consider the two models in Figure 55. In order to construct the first, wire-frame, model a system needs only to be told the coordinates of eight points and the nature of the entities which link them. But the creation of the surfaces which clad the second model involves the system in calculating the position of control points (the nodes of the mesh) before linking them all up to form the surface lattice.

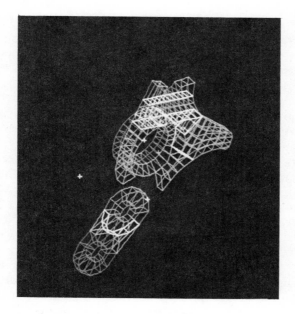

Figure 54. *Finite element mesh.*

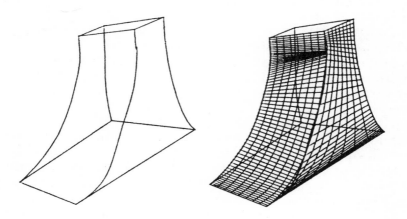

Figure 55. *Moving from wire-frame modelling to surface modelling involves a heavy cost in terms of computer resources. The wire-frame model on the left, for example, can be constructed by the computer given the coordinates of just eight points in its model space and definitions of the twelve entities (8 lines and 4 B-spline curves) which link them. In order to generate the surface mesh the computer must calculate the form of the equation for each of the lines which forms it.*

It follows that, in order to obtain a reasonable level of performance and an acceptable speed of response, a surface modeller requires not only more elaborate programming but also a substantial increment in processing power compared with a wire-frame system. The cost is likely to be correspondingly greater. In considering whether or not this can be justified, it should be recognized that surface modelling not only makes possible applications which would be impossible, or at least scarcely practical, with wire-frame models, it also offers important advantages in terms of convenience and graphical presentation.

Hidden line, or hidden surface, removal for instance becomes a comparatively simple matter. It is no longer necessary for the designer to 'declare' which portions of a model represent a surface. Instead the system itself will assume that all surfaces conceal what lies behind them unless they are specifically declared to be transparent. This is a very important point since surface models in which the hidden surfaces are not eliminated from view rapidly become almost totally unintelligible to the user of the system – though not, of course, to the system itself.

The intelligibility of a model can, with some systems, be further improved by filling in the surface lattices with solid colours and indicating their contours by simulated shadows and reflections. From the observer's point of view, though not the computer's, the resulting images look like solid models. This is, however, a technique which is even more demanding in terms of programming and processing capacity and one which is only available on systems using raster display. Stroke-writing systems, rather obviously, cannot generate blocks of colour or tone.

But, however realistic a surface model may appear, it must always be remembered that the system itself still has a very limited knowledge of the object being modelled. To come back, for example, to the combination of models we used earlier, the cube with the cylinder running through it, a surface modeller will know about their surfaces in a way that a wire-frame modeller cannot, but it will know nothing about the spaces enclosed by those surfaces. Thus, if we look at the surface model shown in Figure 56, it can be seen that the form of the surfaces where the two objects intersect is now well defined. But we still cannot necessarily discover, from the information available in the system, the volume of that part of the cube which is occupied by the cylinder. If, that is to say, the cylinder represented a hole in the cube the system might well be unable to calculate the volume of that hole.

If we require information of this kind, which could be vital to the design of a chemical plant, say, or for determining the size of the 'shot' needed for an injection moulding, then we will have to turn to one or other of the two possible approaches to solid modelling. The first of these to be considered, that based on the notion of 'boundary representation', employs geometrical principles which are developments of those used in wire-frame and surface modellers. But the second approach, based on the

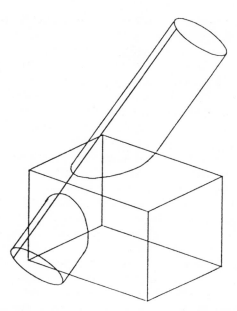

Figure 56. *Once the wire frame is clad with surfaces it becomes possible for the computer to calculate what happens when two surfaces intersect, but it does not necessarily follow that all of the properties of the model can be deduced from a knowledge of its surfaces.*

use of primitive geometrical solids, marks a radical departure in CAD techniques, both in terms of the mathematics which underlie it and in the whole way in which it forces the designer to think about his work.

Solid Modelling I: Boundary Representation

Given that a surface modeller will have a complete knowledge of the positions of the surfaces or boundaries of the model which it contains, it is reasonably easy to see that further programming would allow it to understand them as solids – essentially what is needed is some means of distinguishing a space that is totally bounded by surfaces, and thus 'solid', from one that is not or, to put it slightly differently, the computer must be able to distinguish the inside of a surface from the outside. Boundary representation (or B-rep) achieves this by 'sweeping' the model space with either a line that rotates around an axis or a wire frame; in either case the enclosed spaces are identified and treated as solid.

The great advantage of B-rep is that it is essentially a further continuation of the process which generates first a wire frame and then its surface cladding, this means that a B-rep system can be used in a wide variety of ways; if a simple 2-D model will be adequate for a particular purpose, or if there is no need to go beyond the surface-modelling stage, then a system using B-rep can still handle the job. It is also relatively easy, as is certainly not the case with systems using solid geometry, to derive one sort of model from another – to reconvert a solid model into a wire frame, for example.

But there is a large body of opinion which holds that if you are dealing with simulated solids it makes a good deal of sense to handle them in a way which comes closer to the manipulation of real solids rather than to continue to treat them as pictures or plans of solids. Obviously, a CAD system cannot be equipped with anything like our own sophisticated comprehension of a vast range of solid objects, but it can cope with the primitive forms of solid geometry – cubes, cones, cylinders, pyramids, spheres, etc – and by combining them in a variety of ways it can produce more complex models. This is the basis of the second approach to solid modelling: constructive solid geometry (or CSG).

Solid Modelling II: Constructive Solid Geometry

To understand how a model can be created out of primitive solids, the process upon which CSG is based, we must forget all that has been said so far about the way in which models are constructed by assembling graphical entities and look at the problem from a completely different angle. What, for example, are the solid geometrical elements needed to create the now-familiar object shown on the left of Figure 57?

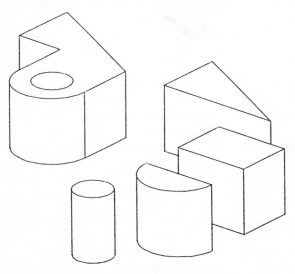

Figure 57. *The complex solid model on the left can be thought of as being composed of the four primitive solids on the right – a wedge, a cube, a cylinder and a half-cylinder, which will itself be created by taking a complete cylinder and removing 50 per cent of it.*

There are four, shown in exploded form on the left of the figure: a cube, a wedge, a cylinder and a half-cylinder. Figure 58 shows the sequence of operations that produces the model. (These are known as

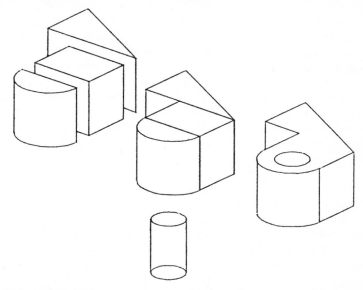

Figure 58. *The steps by which the four primitives on the right of the previous figure are manipulated in order to create a single model.*

Boolean operations since they are based upon the algebra developed by the nineteenth-century mathematician George Boole – the same algebra, incidentally, as that which forms the basis for all computer logic circuitry.) The system is told, in effect, to add the cube to the wedge and the half-cylinder (which will have been created by taking a complete cylinder and 'subtracting' half of it), thus producing a model that contains three of the four primitives. The final step involves 'subtracting' the cylinder from the model in order to create a 'hole'.

Although this may, at first, seem a clumsy and unfamiliar procedure, it does have many advantages. It is often, for instance, a far more rapid way of creating a solid model than B-rep. Since all solids are represented inside the system as combinations of primitives or parts of primitives a task like calculating the volume of a complex space presents few problems; it is just a question of taking the volumes of the constituent primitives and adding or subtracting them as appropriate.

Solid modelling using primitives also has the advantage that it can relatively easily achieve levels of 'realism' and clarity which other systems only attain with difficulty. It is, for example, necessary to build processes like hidden surface removal and simulation of shadow and reflection into the software from scratch, otherwise it would be impossible for the system to represent even the basic primitives on-screen. Given that the system is dealing with entities which are themselves solid, rather than solids which are built up out of many separate entities, it is also rather easier to generate what might be called CAD 'special effects' such as cutaway or exploded views of complex mechanisms. For the business of building up a complex model containing many solid elements mirrors rather closely the methods that might be used with real objects in the real world (it has, aptly, been described as 'computer Lego') and it is correspondingly simple to dismantle the model just as one might dismantle a real-life assembly.

The most important drawback of this approach to solid modelling is that it is difficult for systems which employ it to be interfaced, or even communicate, with those that use the alternative approach. For example, it is not easy to devise a program which can take a model made up out of primitive solids and transform it into a wire-frame model.

There is, none the less, a widespread feeling that the use of CSG will grow over the next ten years. The reasons are not difficult to see. Designers in many different fields, once they have got over the novelty, find that building a shape up out of primitives allows them to concentrate on essentials and to see things that were often obscured by the technical difficulties of constructing solids out of lines. The whole process brings them closer to the actual manufacturing process, and is in some cases almost a literal simulation of it. The parallel between boring a hole in an object and subtracting a model of a cylinder from a model of that object is, after all, very close, and adding, say, a model cogwheel to a model cylinder is very like slipping a real cogwheel onto a real axle.

There are areas, of course, in which modellers of this kind would be worse than useless. There would be little point, for example, in laying out a printed circuit board by these means. But there are many others in which it can bring a new freedom to designers. In fields such as mechanical engineering the representation of objects in terms of the space they occupy rather than the positions of their boundaries encourages designers to tackle problems from an angle that may be novel but is also often fruitful.

It also seems likely that this approach will provide the best basis for many of the developments outlined in Chapter 5. For these depend, in large part, upon CAD systems being able to understand and simulate aspects of reality which are not solely geometrical, and there seems little doubt that models which are built up out of solid entities are 'more like' real objects in several respects than those which are developed from traditional graphical representations.

Summary: Making a Choice of Modelling System

In CAD, as elsewhere in life, the rule that you get what you pay for is generally applicable. All other things being equal, higher-resolution colour display, large memory capacity, powerful processing resources and solid-modelling ability will always cost more than low-resolution monotone display, limited memory and processing power and two- or three-dimensional wire-frame modelling. The question that the would-be purchaser must consider in deciding how far up the scale of modelling abilities it is necessary to go is twofold: 'What can I afford?' and 'What do I need?'

Everyone will be able to determine their own response to the first question and we shall be looking later on in the book at how one should go about answering the second in detail. At this juncture, however, the point to grasp is that there is only one thing more wasteful than buying a system with modelling abilities that exceed your needs – and that is buying one whose abilities do not meet them as completely as possible. For, as we shall be arguing in subsequent chapters, the decision to switch to CAD is likely to achieve its best results when it is implemented at every level and in every department of a company that is either a producer or consumer of geometrical design data. Nothing is worse, therefore, than having a system which can handle some tasks but not others, for this will involve data being taken in and out of the system with much consequent waste of time and effort and almost certainly at the price of losing some of the integrity of that information.

One point, perhaps, does need to be emphasized particularly strongly. As has been explained, 2-D, 2½-D and 3-D wire-frame modellers, surface modellers and B-rep solid modellers can be seen as a series of stepping stones leading from what is little more than an electronic drawing board to a full solid-modelling capacity. It may well be possible, therefore, to move from one stepping stone to the next without changing the entire system. Provided the computer resources are sufficient, for example, a wire-frame modeller may be converted into a surface modeller by making additions to the existing software.

But CSG solid modellers that create their models out of primitive solids are likely to be totally incompatible with other systems. We have, perhaps, said enough to indicate that we believe that, in many applications at least, CSG is the way of the future. But it must also be admitted that, unlike modellers which employ the traditional graphic elements, CSG requires those who use it to change their methods of thought and work rather radically.

Chapter 5
The Software – What CAD Can Do

When you invest in a CAD system the most obvious and tangible return that you get for your money is the hardware – the computer(s), workstations, plotters, etc. But many of the turnkey systems sold under a supplier's own brand name are little more than a collection of standard hardware items available off the shelf from the manufacturers. What gives such systems their distinctive identities, and distinguishes them from rival systems, is not so much the choice of hardware components or even the way in which they are combined with one another, but the range and quality of the software that comes with the system or is available for use with it. In the future some CAD manufacturers may in fact be suppliers only of software; already there are firms who make a virtue out of the fact that their programs are 'hardware independent'.

The most fundamental feature of any system, its modelling ability, is largely a function of the software, as we saw in the previous chapter. The capacity of the hardware is, of course, very significant in that it will determine what sort of software can be used, but the most advanced hardware on the market will be of no use unless it is matched by software capable of exploiting its quality. But within each category – 3-D wire-frame modellers, say, or solid modellers using CSG – there will still be an enormous range of choice, much of it relating to software rather than hardware. In this chapter, therefore, we shall try to provide a comprehensive, if rapid, survey of current CAD programming, concentrating as in the rest of the book on those aspects which are likely to be relevant to the user. There is, for those who wish to pursue the technicalities of the subject, a rapidly growing literature on the mathematical and algorithmic bases of CAD programs, a selection of which is listed in the section 'Select Bibliography' (see page 223).

Almost all CAD programming falls fairly clearly into one of four categories. There are first of all the graphics programs themselves which allow the designer to create a geometrical model, view it on the display screen, transform it in a variety of ways and alter or elaborate it. Then there are analytical programs, designed to test the qualities of a design by simulating the conditions that it would encounter in real life. Next come programs which deal with the creation of manufacturing information; in practice, as we saw in Chapter 1, this means the generation of NC part programs. Finally, there is the programming that handles what might be called the 'administrative' business of the system. Although this type of programming is frequently ignored, it seems to us to be of great, and increasing, importance. No matter how refined are the geometrical models generated within a system, they will be of limited help to the working designer if the files in which they are stored cannot be indexed, cross-referenced and retrieved in a fashion which is flexible and reflects the real relationships between the various elements of a design.

Basic Drafting

The basic task of drawing lines is normally accomplished by the sort of procedures described on pages 72-80, that is, by selecting the type of entity (line, arc, circle, or, in some cases, simple geometrical features such as arrowheads or matching symbols) from a menu, and then specifying size and position by pointing to the display, inputting numerical data, choosing from a range of options on the menu, or some combination of these methods.

So far at least, CAD systems are rarely used for the kind of 'back of an envelope' sketching that designers do at the very earliest stages of a job, so the need to draw freehand on the display screen seldom arises. It is, however, worth noting that the 'rubber-banding' technique in which one end of a line is 'fixed' in position on the screen while the other remains 'attached' to the moving cursor, lends itself to freehand drawing, especially when used with one of the cursor steering input systems. Curiously, perhaps, drawing directly on to the screen with a light pen, or indirectly via a graphics tablet, is a skill that comes less readily to many people.

In most cases, once a model has been created the majority of instructions given by the designer will involve reference to the image on the screen — 'join the end of *this* line to *that* corner', 'draw an arc tangent to both *this* line and *this* one', 'delete *this* line between *here* and *here*', etc. Thus, whatever combination of input devices is employed, there will be a need, as we have explained, for a 'pointer' of some kind. There will also have to be some arrangement whereby the system can acknowledge an instruction and verify to the operator that it has correctly identified the position or entity to which he was pointing. In the case of a cursor steering system or graphics tablet input the screen cursor may 'flash' when the 'tag' button on the input device (or a button on a keyboard or keypad) is pressed, or when the tip of the stylus touches the graphics tablet. Pressing the tag button to indicate that the cursor is now in the desired position may also, in some cases, cause the symbol to be generated to remain in position on the chosen point.

Whatever arrangement is employed it is vital that the system should, to some limited extent, be capable of interpreting an instruction and should not test the designer's skill and patience by being overly literal minded. Nothing is more irritating than a system which refuses to recognize the entity which you are pointing to because the light pen or the cursor is not positioned with pinpoint accuracy. To avoid this problem it is usual for the cursor cross or the tip of the light pen to be surrounded, as it were, with an 'aura of sensitivity', as shown in Figure 59.

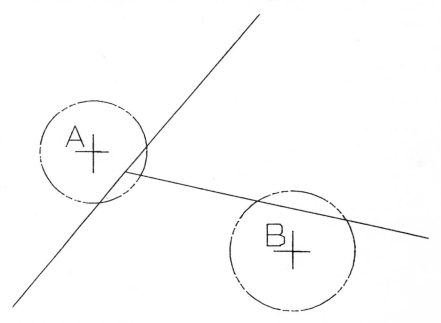

Figure 59. *In order to make the designer's task easier and remove the need for a cursor or light pen to be positioned with total precision, most systems will recognize which entity is being indicated even when the pointer is slightly out of position. The easiest way of visualizing the arrangement is to think of the cursor or the tip of the light pen as being surrounded with an aura of sensitivity so that, in this example, the system can understand that the designer intended to point to the corner, A, where the two lines meet, or to the horizontal line, B.*

It is very important for the newcomer to CAD, who may well find the arrangements for input, whatever they are, clumsy and unfamiliar, to recognize that within a comparatively short period they can become almost second nature to a regular user of the system. There is, after all, nothing inherently 'natural' about using pencil, compasses, set square, protractor, French curves and parallel rulers to construct a plan, and no one should allow themselves to be 'put off' by the fact that drafting with the help of a computer is not like using a drawing board – it may be different, but it is often very much easier!

Macros

Even so, creating a very large model entity by entity can be a long and ted- *ious*
ious business. If you are using pencil and drawing board, of course, it is
more or less unavoidable. But with a CAD system it is perfectly possible
to let the computer handle the repetitive and purely mechanical aspects.
The simplest and most obvious way of achieving this is by the use of
'macros' – standard 'bits' or 'sub-models' which can be added to a model
or subtracted from it by simply issuing a single instruction to the computer.

A circuit designer, for example, will save much wasted time and effort
if he can just tell the computer to insert the standard symbol for a resistor
or a capacitor in a circuit diagram, or to draw the standard components
into the plan for a printed circuit board, in specified positions. Similarly,
an architect will benefit if the CAD system can handle routine chores such
as drawing in the doorways or the washbasins. Many systems will not
only have basic geometrical macros such as triangles, squares, trapez-
iums, etc, already available on their graphics menu, but may also make
use of specialized 'symbol libraries' including anything from standard
pipe joints to hedges or trees. Special-purpose graphics programs, such as
are available for the design of electrical circuits or mechanical engineering, for example, will normally come with menus that incorporate a symbol library appropriate to that particular application.

It is also possible, in virtually every case, for the user to create his own
library of macros, either 'system macros' that will be permanently available to all users of the system, or macros that are specific to one user or
one particular job. The macro is, in fact, simply an additional drawing
that is inserted on command. It is thus a graphical equivalent of the kind
of subroutine that is commonplace in other computer applications. So,
just as a company's accounts department may modify a standard accounting program by adding subroutines that meet their particular needs, a
design department may equip its CAD system with a set of macros that
represent features in everyday use in the department.

Parametrics

The use of macros – standard elements which can be repeated as often as
required – leads fairly naturally to the idea of parametrics – the use of
elements which are, so to speak, 'the same only different'. The most
straightforward example of this technique involves what is known as
copying and mirroring, creating a complete model of a symmetrical
object by drawing one sector and then instructing the computer to generate the remainder by creating copies and mirror images. Figure 60 shows
how this technique simplifies the modelling of objects made up of regular
segments, such as, in this case, a cast wheel.

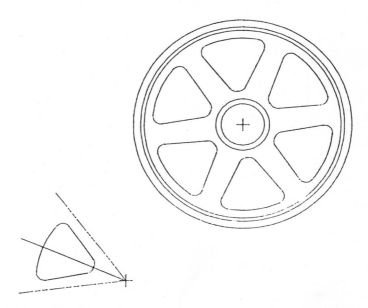

Figure 60. *The use of copying and mirroring can greatly speed up the drawing of symmetrical forms. In this example, just one segment, 60°, of a cast wheel was drawn and the other five segments were added by successive copying operations.*

Other more fundamental parametric techniques depend upon the notion that a single set of graphical instructions can be used to generate a whole range of different, though related, models if one or more parameters are altered. Figure 61, for example, shows a variety of apparently heterogeneous shapes that are all, in fact, variations on a single original. All that has been changed is the distance between the diagonally opposed corners of a z-shaped model.

Models incorporating curves (or, indeed, 3-D models) can, of course, be parametricized as readily as 2-D ones. Figure 62 shows a relatively simple example whereby the 'burgundy' bottle on the left has been transformed into the 'hock' bottle on the right by altering the ratio between four basic parameters.

The ability to alter the form of models by 'stretching' and 'squeezing' them in this way has very practical applications. It might, for example, be of value in the creation of a form such as a tapering aerofoil, or allow many variants of a standard design for an item such as a bolt, perhaps, to be modelled to conform to a set standard by altering certain basic parameters. It can also, clearly, play a large part in simplifying and speeding up the process of trial and error which occupies such a high proportion of many designers' time. If, for example, it turns out that one critical dimension in a complex design must be altered, then rather than having to redraw the model from scratch, the designer can simply tell a parametric

Figure 61. *The use of parametrics allowed all these variations on a z-shaped original to be created by varying just one parameter, the distance between the diagonally opposed corners.*

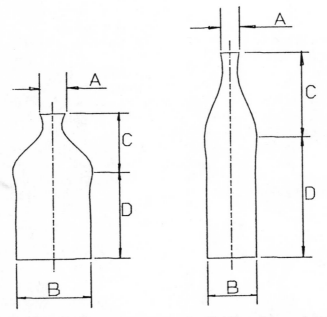

Figure 62. *By changing the ratio between the diameter of the neck (A) and the diameter of the base (B), and the ratio between the length of the neck (C) and the length of the rest of the bottle (D), it was the work of a few moments to 'redraw' the 'burgundy' bottle on the left as the 'hock' bottle on the right.*

modeller to reconstruct its model with a new value for that single distance.

Graphic Conventions

Any 2-D representation of a 3-D object will rely to some extent upon graphic devices or conventions which supply the observer with visual cues as to which interpretation, out of the several that may be possible, is the correct one. We have already looked at two of the more important of these — the removal of hidden lines and depth cueing by varying the intensity of line (pages 92-93). Others which depend upon transforming or sectioning a model will be considered shortly.

But there are other very simple methods of clarifying the presentation of a model, some of which will be familiar from more traditional design methods. For instance, different kinds of weights of line can be used to distinguish one model or part of a model from another, as, even more effectively, can colour. If a model is made up of a number of different layers (see page 86) colour may well be a useful means of distinguishing between them, and it will be valuable whenever two different categories

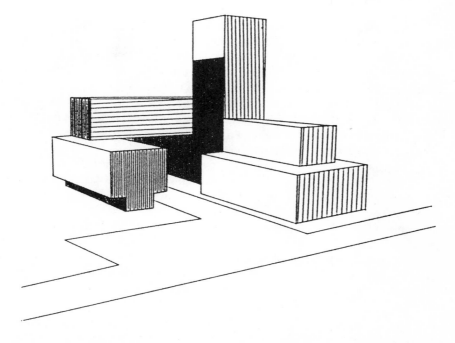

Figure 63. *Even though this complex of buildings has been modelled only as a wire frame, it can be made to appear more realistic by the use of cross-hatching.*

of information are presented at the same time – the geometry of an object, for example, and the tool path needed to create it.

Surfaces, as we saw in Chapter 3, are normally represented by a lattice of lines joining control points, the position of which has been calculated by the system. But even when a system lacks the capacity to determine the form of a surface it may still be possible for the designer to use shading and cross-hatching to provide some visual clues, especially when preparing sketches or drawings for presentational purposes. In stroke-writing display systems, where neither solid blocks of colour nor tone can be employed, cross-hatching is the only means of achieving an illusion of depth in an image. Figure 63 shows a wire-frame building which has been treated in this way.

In the case of raster display there is, of course, scope for far more elaborate graphic devices. Blocks of colour may be used to distinguish models or parts of models, or even the various surfaces on a model, from each other and, given surface- or solid-modelling capacity, the system may be able to simulate the fall of light on the model and the shadows and reflections which it produces.

'Drafting' with Primitive Solids

The difference between building a model up out of graphic entities such as lines and arcs (or even combinations of basic entities such as triangles and squares) and creating a model by adding and subtracting solids such as cubes and cylinders could be compared to the difference between a graphic art like drawing and a plastic one such as sculpture. In both cases, of course, the basic elements of the models exist only as mathematical descriptions inside the CAD system, but whereas models built out of graphic elements could be thought of as 3-D drawings, those which consist of solids are more like simulated objects.

As we saw in the last chapter, the designer uses primitive solids rather like a set of magical building blocks which can not only be added one to another, but can also be subtracted from each other. Every time a new element is added or subtracted it must first be specified in detail; the system must be told what sort of primitive is required (a cone, say), what its dimensions are and how it is to be positioned in the 'model space'. Only then can new solids be created out of various combinations of two or more elements. Figure 64 shows the three basic Boolean operations as they might be applied to two such primitives: a rectangular and a triangular plate.

When models or parts of a model have been created by combining a series of primitive solids they are, in many ways, more 'like' the objects they represent than those which have been constructed out of the conventional graphic entities. In terms of the programming required, they are

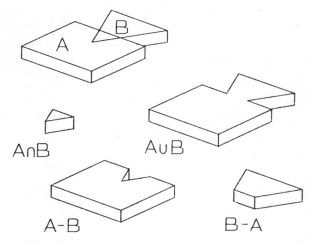

Figure 64. *The three basic Boolean operations applied to two primitive solids: a rectangle (A) and a triangle (B). Union, A∪B, produces the space that contains both A or B; intersection, A∩B, the space that is in both A or B; while A-B, consists of the space in A* minus *the space that is common to B, and B-A, is the space in B* minus *the space that is common to A.*

conceptually easier to manipulate, so that it is possible for a mechanical engineer, say, to dismantle a mechanism on the screen, and modify a component or rearrange the assembly very much as one might fiddle about with a real piece of machinery. Using the same sort of techniques, it is also feasible to generate cutaway and exploded views which clarify what might otherwise be confusing. Colour, rather clearly, is virtually indispensable to this sort of process. A model of a complex mechanism containing a number of separate parts, each intricately shaped, will be totally incomprehensible, even to the person who designed it, unless different colours are used to make each part distinguishable from its neighbours.

Transformations

Although the type of model which a system can construct is, from the user's point of view, its most important single characteristic, its capacity for transforming a model comes a very close second. For unless the designer is able to examine the model from the widest possible range of viewpoints and in the greatest possible detail much of the potential benefit of using a CAD system will be lost. There is, after all, very little point in the system having a detailed knowledge of the model's geometrical properties unless it is capable of 'showing' the designer the aspect that he wants to look at.

The range and variety of transformations that can be achieved are, there-fore, a very significant measure of a system's ability. But it is also very directly related to its size and the amount of processing power and mem-ory space that is available. For transforming a model in order to present a different view of it may well be as demanding, in terms of the computa-tions involved, as creating the model in the first place. Although, in the case of a very powerful system, the observer may get a very vivid impres-sion of a 'real' model rotating before his very eyes, it is important to remember that this is an illusion. There is no model, as such, inside the system, just a mass of geometrical data and a set of programs which allows that data to be translated into a variety of different screen images, each of which represents one view of the model. Transforming a model means, in effect, that the computer must redraw the image from scratch, and in many cases that is a long and complex task.

The simplest kind of transformation, and the only one of any import-ance which can be applied to a basic 2-D model, involves 'panning' and 'zooming' in order to examine a portion of it in close up or to 'stand back' from it. This technique highlights one of the great advantages of CAD. Using traditional methods, a designer may well have to make several drawings of the same object on different scales for different purposes – an architect, for example, routinely supplies detailed drawings to clarify a feature that would not be visible on a large-scale plan. But using CAD it is possible in principle to construct a single model which can then be viewed, as it were, at any level of 'magnification' – so long, that is, as all construction is done to the same absolute tolerance. Figure 65 shows, first, a general, large-scale view of a complex mechanism and then a 'blown-up' view of one small part of it.

In the case of even the most basic systems it will be possible for the designer to zoom in and out in this way, moving into close up perhaps in order to refine a detail, and then zooming out to take a look at the model as a whole, or constructing a large-scale model and then zooming in on one area in order to check that elements do not overlap or interfere with one another.

An incident that occurred with a student a year or two ago amusingly illustrates the point that even though the designer's ability to understand an image, and the system's ability to display it accurately, will vary as the viewpoint zooms in and out, the system's knowledge remains constant. In this case the student concerned complained that, following an instruc-tion to zoom out to display all objects that had been drawn, the screen had simply gone blank. The obvious conclusion seemed to be that there had been a malfunction and that rather than transforming the model the system had lost it altogether. But after a good deal of investigation it became clear that in specifying a new point on the model under construc-tion the student had inadvertently misplaced a decimal point in one of the coordinates, and the blank screen represented the system's best effort to

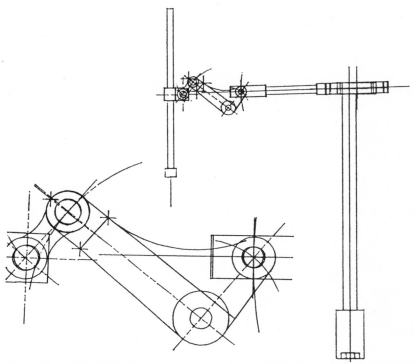

Figure 65. *The image on the top right shows a complete mechanism (the shaft and cam which drive the needle of an industrial sewing machine together with the linkage and bell-crank which transmit the motion). By zooming in on the bell-crank alone, as on the bottom left, it is possible to examine it in detail.*

show both the model and this single point, which were several light years apart! Once retrieved from the far side of the galaxy, as it were, the model proved to be perfectly intact and none the worse for its misadventures.

All 3-D modellers will certainly provide for transformations between the various orthogonal views (the views along the x, y and z axes) and in many cases a system will allow all three to be displayed simultaneously, as in Figure 66. It will also be possible to generate isometric views, such as that shown in Figure 67, since the transformations involved are relatively simple – the orientation of the entities must be calculated but their relative sizes remain constant. Two-and-a-half-dimensional modellers will also usually allow the separate models which exist on different layers to be combined and displayed isometrically, as shown in Figure 43 (page 87).

Many other transformations are, of course, possible. Perspective and oblique, which are commonplace in many applications, may be available, though the former is 'expensive' in computational terms since the relative size of each entity must be recalculated every time the perspective is altered. In the case of very sophisticated and specialized systems containing

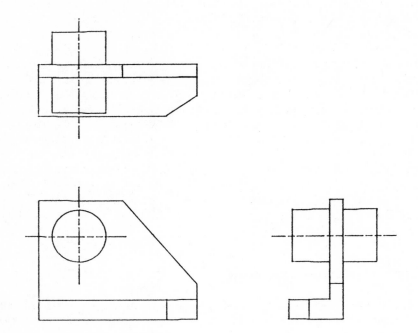

Figure 66. *The three orthogonal views of a model displayed on the same screen, a facility which is standard on several CAD systems.*

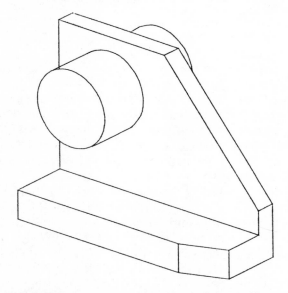

Figure 67. *Isometric view of the object shown in Figure 66.*

their own dynamic graphics processors, extremely complex transformations can be effected in real time so that the observer can fly round a model or, in the case of the flight simulation systems that are now in widespread use, fly *in* one!

In most cases there will be a fairly direct relationship between a system's capacity for transformation and its cost. Remember that the elaborate software required to perform sophisticated transformations will not only occupy a good deal of memory, but will also be useful only if the processing power is available to make it work within a reasonable time – no designer is going to make much use of a program which takes ten minutes to generate a new view. The prospective purchaser must, therefore, carefully weigh up his own present needs (and possible future needs) and, on the one hand, ensure that a system will provide the transformational capacity that is actually required while, on the other hand, avoid being seduced by elegant geometrical tricks which may be fun but are unlikely to be of real use.

Taking Things Apart – Sectioning

Just as it is possible to learn only so much by inspecting the outside of a mechanism or a building, so different views of a model, no matter how various and realistic, may not be enough to achieve total clarity and lack of ambiguity. Three-dimensional modellers will, therefore, be likely to make some provision for generating cross-sections or cutaway views. As we have already pointed out (pages 93-95), such techniques have only limited applications in the case of wire-frame models, but they become extremely important when used in conjunction with surface or solid models.

In order to make use of a cross-sectional view, the designer must be able to specify the cutting plane (normally by 'drawing' it on-screen, as in Figure 68) and the viewpoint that is required. A wire-frame modeller can, of course, only join the exposed 'wire ends' up with straight lines to define the surface exposed by the cutting plane, as in Figure 69, but a surface modeller may be able to eliminate the hidden parts and, on being so instructed, generate cross-hatching or even solid colour to clarify the image further, as in Figure 70.

Where a model is made up of separate parts that have been brought together to form an assembly, the easiest way of clarifying a view may be to remove some outside parts in order to reveal what lies inside, just as in real life one might take the casing off a machine in order to examine its inner workings. Figure 71 shows a model which has been dismantled in this fashion. The technique, along with the rather similar one of exploded drawings, is especially valuable in applications such as the production of manuals and other technical literature. Rather than hiring an artist to

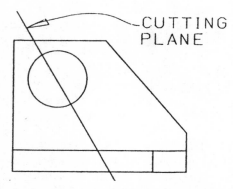

Figure 68. *By drawing a cutting plane across the model as displayed and indicating the viewpoint from which it is desired to look at the resulting cross-section the sort of images shown in Figures 69 and 70 can be created.*

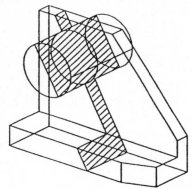

Figure 69. *When a wire-frame model is cross-sectioned what is, in effect, created is another wire frame linking the points at which the cutting plane intersects with the wires of the existing frame.*

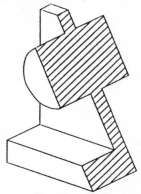

Figure 70. *When a surface or solid model is cross-sectioned it will normally be possible to eliminate the unwanted section of the model in order to leave the newly created surface exposed.*

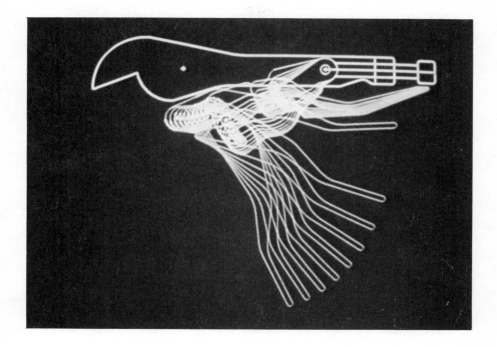

Figure 71. *A model with the 'casing' removed in order to reveal its workings. Here, kinematics have been used to simulate the actual operation of the mechanism.*

draw laboriously an exploded view of a complex assembly, how much easier it is to take the original design and explode it or cut it away.

Putting Things Together – Segmentation and Assembly

For a variety of reasons it is usual to build a large or elaborate model up out of elements or segments which correspond to the parts or components which will be combined to form the real-life object. These elements may, at one extreme, be no more than a separate drawing representing some simple features such as a spot face; at the other extreme they may be complex models in their own right representing, say, a crankcase or an automobile body shell. Sooner or later, however, it is likely to be important to bring two or more segments or combinations of segments together in order to make sure that they will fit together accurately or to check that a mechanism will indeed function in the fashion that has been envisaged.

In the case of a large and complex product such as a motor car or an aeroplane it is likely that different parts of the design will have been produced by different people in different places (or that a design generated in one place will then be transformed into a set of working drawings or a NC part program elsewhere). If CAD is used it can make the business of assembling and checking the disparate elements of a large design incomparably easier.

The example of Boeing's experience (described in detail on pages 147-148) shows the immense gains that can be made when the entire design operation is carefully planned and coordinated from the start. The other side of the coin has been evident recently in the form of strained relationships between large car manufacturers and their subcontractors. The former, naturally enough, now take the use of CAD for granted; equally naturally, in view of their size and resources, they tend to have chosen their own software systems designed to meet their particular needs. The result is that when they invite tenders for the manufacture of new components they would prefer to provide not a set of blueprints but a disc or tape generated by their own CAD system. This is all well and good and commendably up to date, except that the data may be unintelligible to the subcontractor's CAD/CAM equipment, which is likely to be based on one of the standard turnkey systems.

Within a single design department, or a single company, such problems should not arise, though it must be said that there are firms where they are not altogether unknown. But there is a need, none the less, to carefully monitor and control a design job where several different people with different skills and different interests may be working on various aspects simultaneously. The fact that information can be moved around a system with ease – that, for example, a part programmer at one workstation can achieve instant access to data created by a draftsman working at another, or that the chief designer can summon up on his display screen geometrical information that has been generated by half a dozen subordinates working at the other end of the corridor or even many miles away – means that it is extremely important to lay down firm rules. It is fatally easy for modifications to be made without consideration being given to all their implications unless strict protocols are laid down and observed.

Moving Things About – Simulated Operations

It is possible, using CAD, to take things one step beyond the stage of an assembly drawing. The designer can not only check that all the bits fit together correctly, but he can also simulate the actual operation of a mechanism, thus using the mathematical model inside the system as a substitute for the sort of prototype model which might previously have been required.

Kinematic programs range from the relatively simple, such as might be used to simulate the functioning of a basic mechanism like that shown in Figure 72, to the highly elaborate ones used to simulate robot operations on an assembly line.

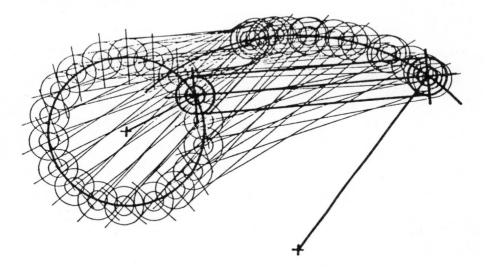

Figure 72. *The operation of an elementary mechanism, in this case a four-bar chain, as simulated on-screen.*

Understandably, the more sophisticated kinematic software is likely to be extremely expensive and, in most cases, to be either commissioned from specialist vendors or developed 'in-house' by large companies with a specific purpose in mind. On the very frontiers of the field is the sort of real-time animation which allows a pilot to engage in a dogfight or land at an airport on the other side of the world without ever leaving the ground. Less spectacular and dramatic, but perhaps of more interest to the average user of CAD, is the software available 'off the shelf' for use with turnkey systems. This may not create realistic illusions of motion, but it does offer the engineer the opportunity to fiddle about and experiment on the screen with ideas that, previously, could be tested only in actual practice.

Automatic Dimensioning

Once a design is complete, there is a mass of additional information to be added to the graphical information contained in the drawings themselves in order to ensure that the presentation is clear, complete and unambiguous. Dimensions and tolerances must be provided; lines and arrows must be added as necessary to indicate which distance is being specified or the exact orientation of a cross-section; cross-hatching or other devices may be required to aid the interpretation of a plan, and so on. The sheer amount of additional work involved using traditional methods can be appreciated by comparing the three-view drawing in Figure 66 (page 116) with the fully dimensioned version in Figure 73. It is easy to see that this sort of work, however straightforward and mechanical, will occupy a large part of any draftsman's working day and involve much painstaking and meticulous labour.

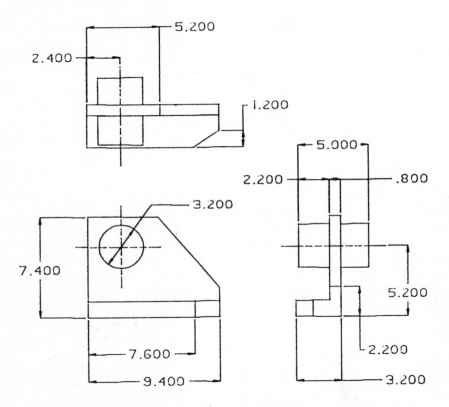

Figure 73. *A finished set of engineering drawings with all dimensions added. Compare this with the same set of three-view drawings shown in Figure 66 on page 116.*

Happily, this is one area where there can be no debate at all about the benefits that accrue from the use of CAD. Virtually every system will be equipped with some basic programming that will allow it to take over a good deal of the donkey work. Even the fact that alphanumerical characters can be typed on the keyboard rather than having to be written out freehand saves much time in many applications. In most well-established fields such as mechanical or electrical engineering, architecture, etc, there are universally recognized standards for the presentation of design information and it is comparatively easy to write programs which incorporate these.

Enthusiasm for the benefits of automatic dimensioning facilities should, however, be tempered by the recognition that the system has no understanding of either the nature of the object represented by the drawing or the dimensions and their function. It is important to remember that it is still the operator's responsibility to decide how, where and to what tolerance a dimension should be set. It is, if anything, even more important for the operator to use his judgement when determining which features should be dimensioned in order to make sure that the resulting description is unambiguous and allows the part to be made within manufacturing tolerances.

If not carefully controlled and used with discretion automatic dimensioning can encourage an attitude of leaving it to the system to insert dimension more or less ad lib. This can lead to very severe problems, as arise from the build-up of tolerances for example, ambiguity, additional constraints and unforeseen difficulties at the assembly stage.

Automatic dimensioning is not, therefore, able simply to take over all the designer's responsibilities in this area; what it can do is to save him a good deal of time and trouble. The system will probably be capable, for example, of automatically aligning labels and captions and of checking that they do not interfere with one another or confuse the plan itself. Standard symbols are likely to be available on the menu and there will be provision for the designer to add to them if and when necessary. Figure 74 shows an engineering drawing incorporating a number of standard features, all of which have been provided from the basic repertoire of a turnkey CAD system. It took no more than a few minutes' work at the graphics terminal to produce this formal presentation of an existing geometrical model, whereas using traditional methods the task might well have occupied a skilled worker for several hours.

The fact that most systems allow different parts or aspects of a model to be stored on different levels can also be extremely advantageous when it comes to presentation. Consider, for example, the three representations of the same object shown in Figure 75: the first shows the geometrical basis of its construction, the second the sort of information that an engineer would require were he planning to incorporate the component in some larger mechanism, the third shows the tool path needed to produce the object. If all the information in the three drawings was combined the result would be confusing, to say the least, and would probably be of little help to anyone. Yet, using CAD, it is perfectly possible to display or generate hard copies of any one of the representations, or of any combination of them, simply by supplying brief instructions to the system.

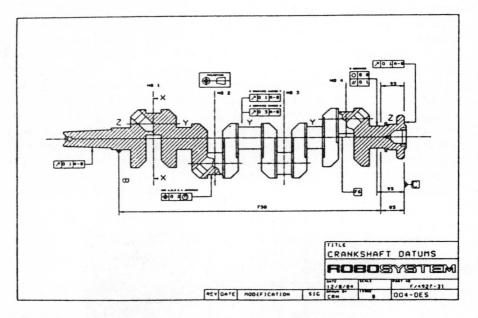

Figure 74. *This engineering drawing, fully dimensioned and containing the full repertoire of standard symbols as well as datum points, etc, was produced within a matter of minutes by an automatic dimensioning program. (Figure reproduced by courtesy of Robocom Ltd.)*

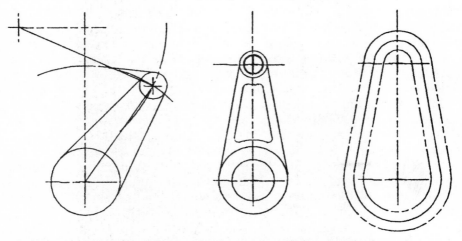

Figure 75. *These three drawings all represent a different sort of information about a single object. The one on the left shows the geometrical basis of the object's construction; in the centre is an engineering drawing of it; on the right is a simulation of the tool path required to machine the object. All three drawings are based on a single geometrical model.*

Testing Things – Analytical Programs

We have already mentioned one type of analysis: that which can be achieved by the use of kinematics. But the analytical software in most general and widespread use is, almost certainly, that employed for finite element analysis. Finite element analysis involves two stages. First the model must be broken down into a series of separate elements by the construction of a grid (see Figure 76). The number and size of the elements will, of course, depend upon the nature of the design and the degree of precision required in the analysis: the smaller the elements the more accurate the analysis – that is until the limit of the computer's capacity is reached and the errors caused by 'rounding' figures upwards or downwards become larger than the numbers themselves.

The second stage involves equipping the computer with software that

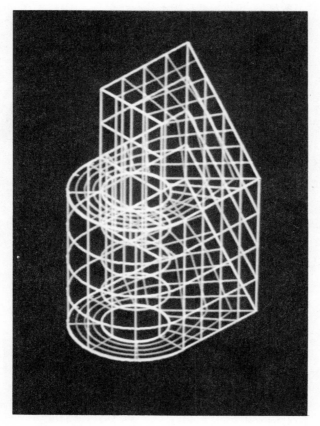

Figure 76. *In this example of a finite elements mesh all the surfaces of the model have been 'meshed'. (Figure reproduced by courtesy of Computervision.)*

allows it to determine some material-related property of each element – typically, its stiffness – and thus to calculate how it will respond to stress. The equilibrium equations for the entire model must then be solved in order to discover where and to what extent distortion will occur under any given set of circumstances. In some cases finite element programming will even simulate, on screen, the result of subjecting a model to a given pattern of stress, see Figure 77.

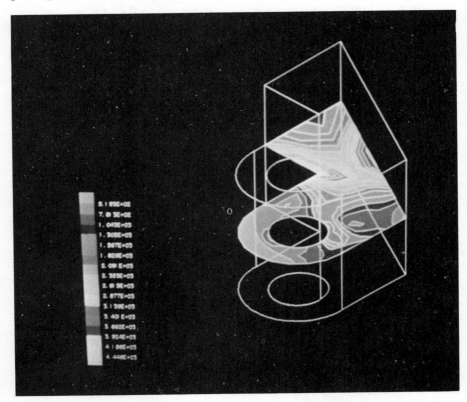

Figure 77. *The stress calculated on the mid-plane of the model shown in Figure 76: the different tones each represent a specified range of stress; on a display screen the areas under the highest stress would appear in red and those that were comparatively unstressed in blue.*

Other analytical software has been developed for particular purposes. One example is 'mouldflow' (see pages 151-152); another is the special-purpose programs which have been developed for use in the design of pipework. Here the main analytical function involves the detection of clashes where two pipe-runs intersect and occupy the same space. Figure 78 shows a model of a complex installation in which the computer has identified two such problems.

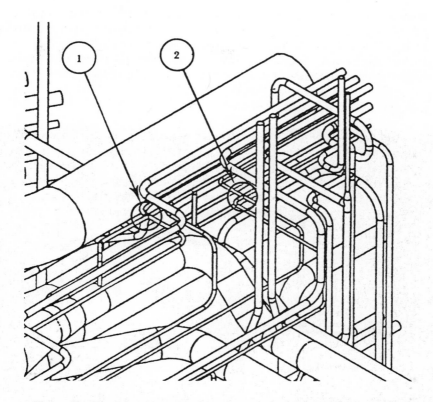

Figure 78. *The Plant Design and Management System (PDMS) software developed by CADcentre Ltd of Cambridge, England is a good example of contemporary analytical programming. This image, part of a much larger model, indicates two locations at which the system has identified 'collisions' between different pipe runs.*

The computer's ability to simulate the behaviour of other machines and mechanical systems, of natural systems such as the weather or the movement of the heavenly bodies, even, some would claim, the behaviour of human beings, makes it likely that the analysis of designs will become of growing importance in CAD. As systems become capable of simulating a wider range of properties – the strength, hardness, conductivity, etc, as well as the geometry of objects – so they will be able to simulate and analyse the behaviour of those objects in a variety of ways. The likely trend of such developments is discussed in the following chapter, but a couple of examples drawn from current practice will show the sort of thing we have in mind.

Designers of printed circuit boards (PCBs) can already test their circuits on a CAD system by using programs which simulate the electrical properties of combinations of electronic components. This means, in effect, that the designer can be sure that a new PCB will do what is required of it without the need to solder a single transistor into position on a prototype. Similar programs are used in the design of silicon chips, but in that case a further refinement is available in the form of other software which simulates the logical behaviour of a circuit containing perhaps 10,000 logic gates.

We are all accustomed by now to the notion that computer simulation may provide a valuable means of reconstructing what happened in an aircraft disaster or predicting what is likely to happen in some sector of the economy. It is also worth remembering that the ancestors of our present computers were designed during the Second World War principally to simulate the trajectories of artillery shells and thus speed up the production of range tables. CAD will, it seems possible, offer the designers of the future the opportunity to simulate the behaviour of many different products under many different conditions and thus vastly simplify the business of 'getting the bugs out' of almost everything from a steam iron to a satellite.

Chapter 6
A Look Ahead

Given the resources that manufacturers are putting into the development of CAD/CAM in terms of both hardware and software, there is no doubt that the next ten years will see the technology advance on many different fronts. The likely nature and direction of these developments are highly relevant to the business of choosing and installing a CAD system today, for the extent to which a business reaps the full benefits of such a step will almost certainly depend upon how readily, rapidly and economically that system can be enlarged or adapted to take advantage of future developments in the field. The purpose of this chapter is, therefore, to try and part the curtains a little and peer ahead into the future.

Towards Standardization?

Perhaps the single most important task for the burgeoning CAD industry, and one it is already limbering up to tackle, is the imposition of a degree of standardization. The present situation, in which different manufacturers have developed different systems independently of one another, is already creating severe problems. It is not unknown, for example, for different divisions of the same large company to find themselves using mutually incompatible CAD systems for different purposes, or different versions of a single program. This can lead to the near-farcical position in which the transfer of information from one computer data base to another involves hundreds of man-hours of tedious work and the non-productive use of expensive computer facilities.

Although the industry is beginning to face up to this problem, individual manufacturers are moving with varying degrees of reluctance — for obvious reasons. Inter-compatibility between different systems will free the customer from the chains which at present bind him tightly to the products of a single supplier. It is the small and medium-sized user who finds himself most at a disadvantage in this situation. Very large companies can afford to devise their own purpose-orientated software, including where necessary programs to ensure efficient communication between different systems. But the firm which purchases a 'turnkey' system, complete with the manufacturer's own software, is likely to find himself locked into a dependence upon that single supplier. He may, in the more extreme cases, find that the task of keeping his system up to date involves the periodic purchase of a whole new package.

This situation, in which the user can become 'hooked' on the products of a single manufacturer, and can only keep pace with advances by means of a series of expensive 'fixes', will almost certainly change with the growth of specialist software houses supplying general-purpose or tailor-made programming that is not exclusive to any single brand of hardware. In the mean time the prospective purchaser, before plumping for any particular system, should try to determine just how far he is tying himself to the apron strings of that particular supplier. The 'clubs' of users which have grown up in order to present the consumer's view to individual manufacturers are one source of information on this question; they may not be privy to the future intentions of their particular supplier, but they will know about their past record. But, in the long run, the adoption of industry-wide standards is the only solution that will be truly satisfactory for the customer.

There is, however, another side to the coin. The imposition of too rigid a set of standards at too early a stage could well act as a drag on future developments. It is already clear that many of the most active 'standardizers' are to be found among the major users of CAD/CAM who see a consolidation at the present stage of development as being very much in their interests. This could lead to a situation in which the standards that are set coalesce not around the highest common multiple of existing CAD/CAM technology but around the lowest common factor. All future advances would then have to be incorporated in an established framework as amendments or extensions of the standards set, which could ultimately lead to more rather than less confusion.

It must also be said that standardization is proving a good deal harder to achieve in practice than was originally anticipated. Already thousands of committee-hours must have been devoted to the subject, but discussions of a protocol governing the transfer of lines and text have not yet been concluded, there is little agreement on one for surfaces, none at all on one for solids and the communication of concepts and functions has yet to be considered!

Horses for Courses: Tailor-made CAD

One trend that is already becoming very evident is the growth and spread of CAD systems designed with specific applications in mind. These may take the form of entire installations, packages of hardware and software purpose-built to meet the needs of architects, for example, or electronic engineers. They may also be offered in the form of specialized 'applications packages', in many cases adaptable for use on a variety of systems.

There is no doubt that as the use of CAD spreads the range and variety of specialized software will widen, and it will be by no means limited to traditional design applications. Already, for example, CAD techniques are being used for tasks as varied as marketing and planning the phasing of large-scale projects.

As an aid to the sales department, CAD can be used not just to give a more realistic impression of what, say, a completed building will look like, it can also be used to help the customer determine and plan his own needs. It's use, for example, has been considered by companies offering kitchen units and equipment direct to the public. The customer would simply provide the dimensions of his or her own kitchen, this information could easily be transformed, by specially written programs, into a 3-D wire-frame model, and with the whole range of available units, appliances, etc, on call as system macros, different arrangements and combinations could be tried out and displayed on the screen. Carried to its logical conclusion, such a system could not only help the sales staff to display the options available, it could also be used to provide 'instant' estimates, and even, when the choice had been made, to generate the sales, delivery and stock-control paperwork. This is far from being a mere pipe-dream. A similar technique is already being used by some firms to help in the marketing of process plant where, on a larger scale, similar problems of combining standard units economically and efficiently within a given space arise.

Applications such as these are, at the moment, exceptional; but there seems little doubt that as experience of CAD and familiarity with its use build up, more and more of them will emerge.

Extending CAM – Computer-aided Everything

The links between CAD and CAM are likely to widen and ramify, as the possibilities of exploiting the geometrical data base created by CAD are explored. The first evidence of such a trend is already emerging in the shape of computer-aided test (CAT) and computer-aided quality control (CAQC). In both instances, of course, use may be made of computer-based techniques that are quite separate from and unrelated to the CAD system on which the products were designed. However, there are also many cases in which the relationship is very direct.

Modern machining centres are frequently equipped with sensing probes which can be automatically loaded by the tool-changing mechanism and used, at predetermined intervals, to check the accuracy of the processes so far completed. Such pauses for inspection can not only be specified in the NC part program, the dimensions and tolerances to be checked can also be included.

Similarly, if piece parts are to be gauged or otherwise tested as they pass through the manufacturing stages, the necessary information may form part of the data base of a DNC or CNC system.

An excellent, if less dramatic, example cropped up in a recent consultancy job. The problem arose in the toolroom of a major manufacturing firm, responsible for producing a wide range of moulds and other tools that had to be machined with extreme precision. The ball gauges used by inspectors to verify that the finished items met their specifications had to be individually calibrated, and each inspector had to take account of the allowance for his particular gauge. Since the shop was principally engaged in producing one-off tools, or, at the most, small batches of identical items, inspectors were constantly demanding that the drawing office translate the dimensions to be checked, as given on the specifications, into the actual figures they were to look for after allowing for the variation established for their own gauge. Once the problem had been spotted the solution was clear: allow the inspectors access to the CAD files concerned and provide them with a limited capacity to interrogate the system and to record the information relating to their own gauge – they could then identify the particular dimensions they were concerned with, input the calibration of their own gauge and be supplied almost instantly with the figures they had to look for when they read their gauges in the course of an inspection.

It is not necessary to plunge into the complexities of group technology or computer-integrated manufacture to find many such comparatively trivial instances in which the availability of the original geometrical information generated in the course of design can be used to improve, speed up or simplify functions further down the line.

It may indeed emerge that CAD/CAM is often its own worst enemy in that a great deal of the debate about its future is conducted at the rarified level of new 'concepts' or 'manufacturing philosophies' which would, if put into effect, require wholesale change. It seems possible, and in many ways preferable, that new uses and applications could be found *ad hoc* and improvised on a piecemeal basis in the course of coping with run-of-the-mill problems. The possibility of this occurring will be greatly enhanced as CAD becomes 'demystified' and all concerned, from the shop-floor upwards, realize its potential value as a source of information rather than treating it as an arcane discipline impenetrable to those lacking a master's degree in engineering.

Building-in More Knowledge – Expert Systems

The biggest single advance that is foreseeable, and the one to which we shall devote the most attention, centres upon the notion of making CAD systems more intelligent. In this context, we are not talking about increasing or further devolving their computing powers, but about the possibility of equipping them with a knowledge of the real world which approximates more closely to that of a human worker – be he a graduate engineer or a machine operator.

As we have already stressed, existing CAD systems are almost pathetically ignorant about the physical attributes of the objects which their models represent. To be sure, the computer can be told that the contents of a particular file should be labelled 'engine support bracket', that 'steel casting' should be entered against the code for 'material' and that other information, such as the machines and tools required to manufacture the bracket, and cross-references to other files, say those relating to engine and chassis, should be appended. But all this amounts to no more than the provision of an indexing system that will make the user's life easier; it does not mean that the computer 'knows' anything about the properties of steel castings, 'understands' the function of a support bracket, or has any comprehension of the processes involved in producing one. It would not demur if the title of the file was changed to 'wedding ring' and its contents revised to specify the use of gold as a material and the use of alchemy as a manufacturing technique.

Most of the hopes for achieving a breakthrough in this direction over the next five to ten years depend upon the relatively novel branch of programming concerned with building 'expert (or knowledge-based) systems'. Originating with research work in artificial intelligence, expert systems are designed to enable computers to handle the sort of information involved in any area of professional human expertise. The earliest applications have been in pure science, medicine, geology and law. In the case of CAD/CAM, expert systems offer a prospect of equipping a computer with the means – in terms both of a data base and of reasoning powers – to think more like a human designer.

For example, a designer or NC programmer will currently have to refer to standard printed reference works to check, say, the tensile strength of a standard steel bar or the appropriate feed rate for milling aluminium. It is quite feasible to build such knowledge into a data base of the type used in expert systems and to arrange that the computer draw upon it when and as appropriate, thus 'automating' one of the more tedious and mechanical aspects of life in any design office.

Similar methods would allow the actual organization and retrieval of design files to be improved. At present, such information, if 'computerized', will be arranged in rigidly hierarchical fashion; in order to relate the plans for one minor component to those of another, for example, it will be necessary to follow the structure of the file 'up' through the level of subassemblies and assemblies to the complete machine and then 'down' again. It is not possible to simply say, in effect, 'show me the studs that screw into that casting and fit into the holes in this one'. Expert systems, with their ability to think 'intelligently' about their 'knowledge base', may take CAD a step nearer to coping with such challenges.

More ambitiously, perhaps, it is hoped that expert systems may allow computers to be provided with some degree of what in human terms would be considered basic common sense. A good deal of our own ability to understand and make sense of graphical descriptions depends upon the fact that we all share a certain basic knowledge of the way the world works – we know that gravity makes things fall downwards, that nuts screw on to bolts, that buttons are there to be pushed and levers are there to be pulled, that steel is hard and rubber is elastic, etc. We human beings, of course, take such knowledge for granted much of the time and it is only when we use CAD (or other computer-based information systems) that we realize how much we depend upon it and how absurdly 'stupid' and 'literal minded' the lack of it will make an otherwise 'intelligent' machine.

Because expert systems are designed to incorporate not just the knowledge which a human expert possesses but also the principles and 'rules of thumb' which he uses when he applies that knowledge, there is some hope that they may be able to 'capture' the sort of day-to-day expertise which every designer relies upon. At the moment of course most CAD systems cannot even understand ideas as basic as 'solidity' or 'hollowness', 'flexibility' or 'rigidity', etc. It is likely, therefore, to be some time before we see systems with sufficient expertise to point out to a designer that, say, he has forgotten to allow space for the nut which must surely be required on the end of that bolt, or to the architect that there just will not be room for the door to open fully if the bath is positioned there. But such things may come and, in the meantime, there are certain kinds of expertise which human beings find tiresome to cope with but which computers can handle relatively easily.

Any designer working in mechanical or structural engineering, for example, will regularly need to consult the standard reference works which contain data on the strengths of different materials, the specifications of standard components and so forth. If this sort of information could be built into a CAD applications program it would not only save much time and trouble, it would also allow the system to monitor a design as it progressed and sound the alarm if, for example, a component was going to be subjected to stress that exceeded its design specification, or if guidelines on, say, the extent to which structural elements of a building could be pierced were in danger of being exceeded.

It is likely that most expert systems developed for CAD purposes will be specific to one particular area of application. This is natural enough, as any expert is likely to be expert in only one field. A couple of examples of programs already in use or under development suggest the way in which things are likely to go.

One of the very first expert systems to find practical application outside the laboratory was that devised by the Digital Equipment Corporation to assist in configuring its VAX computers to meet the requirements of individual customers. Although the VAX range consists of a number of standard CPUs (which are found, in fact, at the heart of many turnkey CAD installations) every order will differ considerably from others in that customers can specify their requirement from a variety of options covering the amount of memory, the provision of interfaces, the way in which external storage is to be accessed, etc. In every case, therefore, a different combination of up to 400 components all interconnected in different ways will have to be fitted into the same cabinet. It is this task which has been handed over to the expert system, which can apply the 800 or so rules involved without being fazed by the combinatorial complexities of the problem.

The second example also comes from the world of computer hardware. Today, the use of CAD in the design of silicon chips is a virtual necessity. Simply laying out the circuitry for a chip which may have up to a million separate electronic elements is a task that, by its very scale, has become impractical for human beings to undertake unaided. In fact, however, all the indications are that within the next few years the whole business will have become highly automated. The use of 'silicon compilers', programs which, once given a specification of the logical functions that are required, will create the necessary circuit design, is now commonplace when relatively simple circuitry is involved. In the future, if some of the plans currently being mooted for chips with tens of millions of elements or the integration of whole 'wafers' of chips into single functional elements are to be implemented, then it seems likely that there will be little choice but to construct a program in which the bulk of the design work can be delegated to a computer.

A more experimental application, in the general engineering field, involves the use of an expert system to create a 3-D model from a standard 2-D set of drawings. Here, rather obviously, the program will have to capture some of the common-sense rules which we ourselves use in order to interpret such information. It may, however, prove to be the case that it will be easier to rework the model in three dimensions from scratch.

There is also a strong likelihood of expert systems playing a part in CAM development. Another experimental system is being used to plan the optimum sequence of machining operations needed to produce a part with a given geometry – the output being in the form of an NC part program which, in theory at least, will produce the desired result in the most economical possible fashion. Group technology, FMS and CIM are all fields which, in principle, look likely to benefit from the work being done with expert systems. For in every case use is continually being made of expertise of one kind and another to solve problems such as: Is it necessary to drill a pilot hole and then enlarge it under these circumstances?, Can this part program written when the material being cut was steel be modified now that the specification has been altered to use titanium?, Can this shape be produced by a vertical milling machine?, and so on.

Trends (and Limitations) in Hardware Development

In our view, it is unlikely that advances in hardware design over the next ten years will affect CAD so dramatically as the sort of software developments which we have been discussing. The most important single theme will almost certainly be the growing flexibility with which computer power can be deployed within a system. We have already seen (in Chapter 3) the erosion of the existing distinction between systems which centralize their intelligence in a single host computer and those which distribute it around a number of more or less autonomous workstations. This trend will continue, as microcomputers grow ever more powerful and networking techniques become ever more sophisticated, with the result that it will be far easier for the prospective CAD user to start with a relatively small system and build it up as the need arises.

The continuing improvements in computer hardware which are expected to take place over the next ten years – especially increased processing speeds, the general adoption of 32-bit architecture and, perhaps, the arrival of the first general-purpose parallel processors – will undoubtedly have benefits for CAD, but are unlikely to revolutionize it. Existing systems frequently have more than adequate computer power; the need is for hardware (networking arrangements) and software (operating systems) which allow it to be exploited more easily and flexibly by a range of users.

There is, however, one intriguing possibility currently just looming over the technological horizon that is worth mentioning. The idea is that some routine software may actually be replaced by hardware in the form of special-purpose chips which, when slotted into a standard microcomputer, say, transform it into a graphics terminal. The industry abounds with rumours that major computer manufacturers have succeeded, or are about to succeed, in implementing an entire graphics program on a single chip. Were this to be achieved, it is likely that the main impact would be at the bottom end of the market; for the most obvious advantage of replacing software by silicon is that it would free memory space, thus increasing the CAD potential of the average business micro or even personal computer quite dramatically.

But there is also the possibility, even more remote at present, that special-purpose CAD programming (a pipework applications program, say, or the NC postprogramming instructions for a particular machine tool) may also become available in the form of 'plug-in' chips which can be added on to a terminal as and when required. Again, the advantage would be that memory capacity at the centre of the system would be freed and the need to shift programs in and out of local memory would be eliminated.

Although such innovations may allow the computations involved in CAD to be handled faster and more efficiently, their impact on the average user is likely to be marginal. Most of the improvements that will make the designer's life easier and his work more effective – a wider range of applications, more 'user friendliness', better dialogue techniques, etc – are likely to come from the programmers rather than the engineers.

Nor, as things stand at present, is there any sign of imminent developments in input or display technology. All three of the main input systems (light pen, graphics tablet and cursor steering) will no doubt continue to have their adherents and manufacturers will doubtless come up with further variations on all three themes; but there is no indication that anyone is contemplating radical innovation. Indeed, even the last such novelty, the mouse, can now be seen to be little more than an existing device, the tracker ball, turned upside-down!

In the long-run many hopes have been pinned on the possibility of voice input – there are obvious attractions in the idea of simply 'telling the computer what to do'. But our own view is that it will be many years before voice recognition systems achieve the necessary level of sophistication and reliability – the prospect loses much of its charm if each operator has to provide the machine with a sample of every command in his natural voice before it will respond to his instructions. There is also the consideration that being able to issue verbal commands will not eliminate the need to 'point' at the display. Moreover, given that any foreseeable system would involve commands being provided in some standard form, there is much to be said for retaining the present arrangement which has the advantage of imposing the discipline upon which effective man-machine communication depends. Not until the twenty-first century are computers likely to be able to understand and interpret the sort of natural, informal language which designers and others use to communicate with one another.

Except in the case of a handful of specialized applications, it seems probable that raster display will become almost universal. But this is a reflection of the fact that, as a general-purpose system, raster display has more to offer most people than stroke writing, rather than because of the prospect of any spectacular improvements in raster technology. In fact it seems that the most advanced systems currently available have probably gone about as far in terms of colour quality and resolution as it is possible to go. It may well be, however, that the cost of high-quality display will fall substantially over the next few years so that the standard of the average turnkey system will improve quite markedly.

One category of hardware remains: that concerned with the creation of hard copies. Here there are signs that significant changes may be in the wind. There is a growing tendency for plotters to have some degree of intelligence, thus allowing an entire model to be loaded from the graphics computer and opening up the possibility that the production of hard copies may be a function that can be cut loose from the overall system and handled by an independent unit into which the geometrical information is loaded on disc or tape.

The COM (Computer on Microfilm) systems recently launched on to the market suggest that one day quite soon plans on paper may even be eliminated from the drawing office altogether to be replaced by electronic media and, in this case, ordinary 35 mm transparencies. The manufacturers have essentially 'shrunk' the orthodox plotting mechanism so that it uses needle-sharp laser beams to draw on the minuscule photographic film without sacrificing precision or definition. What has been gained is speed. Instead of taking a matter of hours, as may be the case when an orthodox pen plotter tackles a single complex model, the production of a permanent copy occupies only a matter of minutes – and rather than ending up with a bulky and awkward roll of oversized paper, you end up with a neat transparency that can be put in a projector, reproduced on any scale or in any form that is convenient, or popped in an envelope and posted anywhere in the world for the price of an ice-cream.

New Roles for CAD

Computer systems of several varieties are already indispensable to any modern business that aspires to grow beyond the cottage industry scale. As those systems become able to communicate more freely with one another, so CAD (and CAM for that matter) will cease to exist in splendid isolation and will become part of an overall control and management system. Group technology and computer-aided test and quality control already suggest ways in which purely geometrical information generated in the course of the design process is likely to be put to use in areas far removed from the designer's traditional province.

In the future it is perfectly possible that a company's CAD system may become a sort of central information bank in which material on a host of topics from inventory levels to costings to promotional publications, all of which is related in one way or another to the design of products, is stored. Today, the chief designer can go to his CAD terminal and summon up a three-view drawing of the 'widget mark II (modification 5)'. Soon the chief buyer will be able to access the system equally easily to enquire what purchases will be needed if a batch of 250 widgets is to be manufactured. Similarly, the sales manager will be able to get an up-to-date costing for that batch of widgets within seconds, together with information on the volume and weight of the goods, which will allow a shipping price to be calculated. Or the marketing department will be able to obtain a cutaway picture of a widget for use in a leaflet, or a kinematic sequence of a widget in action to provide the basis of a promotional video.

Ideas like this are not especially far-fetched in terms of the technology required – much of it is already in place. What is overdue is a recognition of the fact that for many purposes the design of a product (as stored in a CAD system's memory) is as useful as the product itself. We have already

mentioned, for example, the fact that production engineers can use a geometrical description of a product in order to generate geometrical descriptions of the equipment that will be needed to manufacture or assemble
that product, and we have outlined how the salesperson of the future may
use a CAD model as the basis of a sales presentation or as a means of
demonstrating the range of options open to a customer.

There is, it seems to us, a very direct parallel to be drawn between the
impact which 'computerization' of functions like accountancy, credit
control and inventory had on management during the 1960s and 1970s
and the effect which CAD is likely to have upon management in the 1980s
and 1990s. In both cases, a mass of information which was previously
difficult and expensive to obtain, and often unintelligible to any but
specialist staff, can suddenly be made, thanks to the computer, easily,
rapidly and economically accessible.

Exactly how and for what purposes managers will use this new freedom
to access geometrical information is as yet not totally clear, but there can
be little doubt that, with the technology in place, many novel and unexpected opportunities for using it will be found. Just one example which
we recently came across may serve to make the point.

A major manufacturing company had decided to modernize one of its
largest plants, installing new machines to handle a number of large-scale
processes and rearranging the layout of the factory at the same time. In
order that production should suffer as little as possible, it was hoped that
the modernization programme could be handled in phases, with only one
section of the plant out of commission at a time. To the executive placed
in charge of planning the project, the prospects looked nightmarish. Not
only did the physical aspects – the removal of old machinery and the
installation of new, the repartitioning of the floor space, etc – have to
follow a complex critical path, so that no phase involved the undoing of
work that had been completed in a previous phase, but this critical path
had to be reconciled with another one that related to the maintenance of
manufacturing processes at as high a level as possible. Stocks of components had to be built up before a section of the plant was decommissioned
that were sufficient to supply needs until the replacement plant was in
place; staff and facilities had to be moved around to temporary locations
to make room for construction work, and so on.

Happily, it turned out that the company's CAD system offered an ideal
means of coping with this task. A series of separate 2-D plans of the factory space, each representing the situation on completion of one phase of
the work, were created as separate layers of a single model. This allowed
the planning staff to see clearly what the position would look like at each
stage, what demolition and reconstruction work would have been completed, what old machinery would have gone, and what new machinery
would have been installed. Different layers of the model could be juxtaposed on-screen in order to identify snags that might otherwise only have

come to light when work was actually in progress. In one case, for instance, it emerged that if a new wall was built at one stage to enclose a completed section of the factory, then it would have to be demolished at the next in order to allow passage of a large piece of machinery which could not come in by any other route.

By using the facility of the CAD system to simulate the entire project, the firm concerned reckoned that hundreds of thousands of pounds were trimmed off the budget, with as much money again being saved by maintaining production when it would otherwise have been lost as the result of unforeseen dislocations.

This was, of course, an exceptional episode. No company rebuilds its factories on a regular basis. But the moral to be drawn from it is none the less very pertinent: we have barely begun to explore the full range of uses to which CAD can be put, and we will only make the most of the opportunities if we view CAD not as a specialized tool but as a facility for which a hundred and one uses may be found in a business that keeps its mind open.

Near Relations: Computer Graphics and Simulators

One of the obvious ways in which CAD is likely to develop will involve linking up with other computer applications that make use of graphic displays. Much business software already makes extensive use of graphic techniques – to generate pie charts or bar graphs, for instance – and even computer game playing may now employ sophisticated graphic techniques. As a result business and even personal microcomputers now incorporate relatively advanced display technology, and applications programs which allow this to be put to use in CAD work are now starting to come on to the market.

Computer graphics, once the province of those fortunate enough to have access to highly expensive and specialized software and a mainframe computer on which to run it, is also starting to move 'down market'. It will naturally be a long time, if ever, before the most advanced techniques can be practised outside the laboratory. Some, for example, may still require hours of computation on the part of a super-computer in order to generate a single image. But programming such as that used to drive flight simulators, which was the stuff of science fiction ten years ago, is already available in the computer stores, so who knows what the next ten years may hold?

The point that concerns us here, however, is that much of this specialist software, whether it is being used to train real fighter pilots, to create imaginary objects in imaginary universes (like the famous X-wing fighter in *Star Wars*, which was in its day the most elaborate image ever created by computer simulation) or to provide fun for the kids on a rainy day, is very closely linked with the methodology employed in CAD. It may, therefore, not be long before a designer produces a very realistic 'TV picture' of, say, a new piece of packaging or even 'flies' a new aeroplane without getting up from the CAD terminal.

The time may be coming, in short, when actually making and operating something will be little more than a formality since not only the appearance and structure of the object but also the processes involved in manufacturing it and the way in which it works under a whole range of circumstances will have been simulated with great precision. Already, in areas as diverse as weather forecasting (where a 'predictive sequence' based upon computer simulation is now a feature of every TV bulletin) and movie making, the computer is demonstrating its uncanny power to show us what reality will, or might, look like. CAD, it seems likely, will before long make it possible for the designer to present an idea with a degree of realism that will revolutionize the relationship between the creator and his employers, customers or audience.

Chapter 7
Justifying CAD/CAM

Like any other technological innovation, the introduction of CAD/CAM into a business can only be justified by improved performance and, ultimately, higher profits. But measuring the benefits that will accompany a switch to CAD/CAM, and judging where, when and how they are likely to accrue is by no means straightforward.

The Fallacy of Productivity

The yardstick which is most easily and commonly applied is also, in our view, the one most likely to mislead; regrettably, it is the one which has found most favour with the manufacturers of CAD equipment. The assumption which underlies it is that a designer's productivity can be gauged by the number of drawings he produces. A drawing is, after all, the most tangible output of a traditional drawing office so it seems not unreasonable to assume that any tool which will allow the same number of designers to produce more drawings in the same amount of time, or the same number of drawings in less time, must by definition be an improvement.

Unfortunately, however, it is far from being the case that more drawings are equivalent to more, or better, design. On the contrary, more drawings may not only mean that unnecessary work is being done in the design office, or that necessary work is being done inefficiently; they may also result in unnecessary work being done, or necessary work being done inefficiently, at the manufacturing stage. It would in many ways make far more sense to turn the argument on its head and justify CAD on the grounds that it allows designers to do more or better work while producing fewer drawings.

It must also be borne in mind that, even measured on this crude and unsatisfactory basis, the gains made in productivity will vary enormously from one application to another. A 1982 survey by Lipchin recorded gains ranging from over 1700 per cent in the design of integrated circuits to a mere 140 per cent in detailed mechanical drafting. Given that another survey at about the same time suggested that the expense involved in purchasing a medium-sized CAD system could not be justified unless the productivity increase was at least 200 per cent (assuming single-shift utilization) the latter figure is not quite so impressive as it might seem at first sight.

Some fairly representative figures were produced by a recent study undertaken by the Special Engineering Department of Brunel University on behalf of a major British manufacturing company which was considering the introduction of CAD/CAM techniques in its factory. Most of the design work that was involved related to the production of specialized moulds and other tools used in the manufacture of consumer goods and packaging materials and was thus typical of that involved in general and precision engineering where items are to be manufactured either as one-offs or in small batches.

Allowing for the inevitable problems caused by retraining and the debugging of an unfamiliar system, the Brunel study indicated that existing rates of productivity would not be improved upon during the first year – at best they would be maintained. Given that the installation would involve no redundancies among design office staff, it was clear that, after the first 12 months, the project would show a substantial deficit. In the second year, it was estimated that the productivity ratio between the new methods and the old could rise from 1:1 to 2:1, and in the long term it was thought likely that the improvement would stabilize at a figure of at least 3:1. On this basis, the investment would have been recovered in just under three years.

Not-so-simple Arithmetic

But it is worth dwelling on the productivity question a little longer, because detailed study typically reveals a whole range of benefits which were not readily foreseeable and which may never be strictly quantifiable.

For instance, among the wide range of machine tools available in one plant examined was a large and expensive flat-bed machine designed to guide a cutting tool along a path which had to be drawn out and reproduced on a specially prepared film. When the operations of the plant were analysed it became apparent that this costly piece of capital equipment was dramatically underutilized – in fact designers were often obliged to go to considerable lengths to solve their manufacturing problems by the use of alternative and less appropriate methods. It was easy to appreciate their dilemma. The tool paths for the machine had to be drawn on a very large scale and sent away to be reduced and processed on to film, a job that might take up to two weeks. Until the film came back and the machine was actually set to cutting metal there was no way of checking the tool path for accuracy or establishing that the finished product would meet specifications. If and when snags emerged at the metal-cutting stage (which could easily happen) it was necessary to begin the whole time-consuming process all over again.

Enter CAD, in the form of facilities available to the company. Not only could the tool path now be prepared rapidly and relatively painlessly using standard graphics software, it could also be checked by simulating the cutting operations on screen and, lastly, the data produced could be plotted directly on to film within a few hours. Thus, if there were still problems (and the fact that the tool path had now been visually verified reduced the chances of this dramatically) the minimum amount of time was wasted in putting them right and obtaining a new piece of film.

As a result, the potential productivity of the tool in question soared almost overnight. Instead of using it only as a last resort, designers could begin to make use of it in a whole range of applications for which they would never previously have considered it.

Taking another instance of the significant but unquantifiable benefits which may come along with the introduction of CAD/CAM. At another plant it was not unknown for the geometrical calculations involved in the design of complex shapes to occupy a skilled designer for up to eight weeks at a time. Those same calculations could be handled by a CAD system in a mere eight hours. But the real gains went far beyond the saving of time and trouble. Experience had shown that a calculation on this scale, involving many variables, was a tricky business in the hands of even the most painstaking and careful mathematician. There was a tendency, therefore, once an answer had been arrived at, for the head designer to set someone else to work on double-checking it. All too often this process produced a second, different answer and the head designer was left contemplating two equations, unconvinced that either of them was correct. The CAD system not only produced an answer in one-fortieth of the time, it also produced an answer which could be confidently assumed to be free of errors of calculation or mathematical slips.

One final example points to the fact that some of the advantages conferred by CAD cannot be measured by comparing the new with the old, simply because there are occasions upon which the new can achieve things that were beyond the scope of the old. Whilst the study was being concluded the client company was involved in bidding for an extremely important and profitable contract. Essentially, winning the contract turned upon the company's ability to design a component that achieved both a set degree of strength and a precisely prescribed weight. The object in question was comparatively small and success or failure depended, literally, upon whether or not small quantities of material could be trimmed from the profile without destroying its strength and integrity.

Using traditional drafting methods, and relying upon years of accumulated skill and experience, the designers produced what they reckoned to be an appropriate design, so finely judged that part of the vital component would be separated by only a few thousandths of an inch from other components – if it was any larger, they estimated, there was a danger that the parts would no longer fit into the space provided. Yet, it seemed, a competitor had done better, and had somehow, somewhere managed to squeeze in that vital fraction of additional material that made the difference between success and failure.

In a last desperate effort to rescue the situation, the existing design was transferred on to the university's CAD system and, lo and behold, it became clear that there was room to change the shape by adding a little bit here and a little bit there until the necessary amount of weight had been achieved.

We have dwelt upon these examples at some length in order to show that justifying the use of CAD/CAM may involve far more than simply counting the output of drawings from the design department, or the man-hours spent on design work. In many cases, indeed, the easily measurable gains in productivity may be more apparent than real.

In the case of another new CAD installation, the preliminary studies suggested that considerable staff savings could be made once the CAD system was up and running. It was suggested, for example, that the firm concerned, which specialized in the design and manufacture of customized engineering equipment, could dispense with the services of one of its two NC part programmers – an economy that, over a two-year period, would make a substantial contribution towards recovering the cost of the CAD equipment. But, in practice, how would the company manage when the remaining part programmer was on holiday, ill or absent for other reasons? The answer, of course, was that it could not contemplate putting its production schedules constantly at risk in this way. The solution, it quickly became apparent, was not to seek ways of replacing men with machines but to find ways in which men *and* machines might, between them, achieve more or better results than men had been able to manage unaided.

A case study from the architectural field reinforces this point. The job in question was a major one: the design of a large new hospital complex on the outskirts of London. As with all multi-storey buildings of this kind, the arrangement of wiring and piping posed major problems for the architects – not because the job of laying them out is particularly difficult in itself, but because the task of fitting many different systems into a common, confined space rapidly becomes nightmarish in its complexity. Realistically, architects have become reconciled to the idea that it is impossible in the time and with the resources available to achieve an optimum solution; they simply accept that there will be wasted space in the service ducts and beneath the floors.

It came, therefore, as something of a revelation to the hospital architects to discover that, by using CAD and producing a standard layout which could be duplicated on each floor and which made the best possible use of the space, the overall height of the hospital could be reduced by the equivalent of an entire floor, leading to a major saving in constructional costs.

As all these examples indicate, the prospective purchaser who looks upon a CAD system merely as a tool which will enable designers to perform their present functions faster or more efficiently may well be missing the point. In some cases, paradoxically, the real benefits of CAD may only be realized if it is accepted that the new techniques will take more time and cost more money. At least one major corporation, the Boeing aircraft company has found this to be the case and it is likely, we suspect, that many other companies are making similar gains without perhaps quantifying or even recognizing them.

Saving Waste – Consistency of Information

The example of Boeing is interesting, if only because the company's investment in CAD is on a heroic scale, involving a thousand workstations, split between 87 locations in the Seattle area and a private microwave communications system to keep them all in touch with each other and with the Cray super-computers which crunch the vast accumulations of numbers associated with the design of a modern aircraft.

According to Bill Beeby, the Director of Computer Systems in Boeing's Engineering Division, the key advantage the company has derived from switching its design operations over to CAD is that all the geometrical information relating to the design of a new aircraft is now created, stored and employed in a convenient, consistent fashion. He gave an idea of what this means to a firm like Boeing by comparing the task of designing the 757 (the first Boeing plane to be designed on CAD from start to finish) with that of designing one of its predecessors, the 747. The actual design process, he pointed out, had been both longer and more costly in the case of the 757, but, once the project moved out of the design office and on to the factory floor, that additional investment began to pay off in a spectacular fashion.

It is worth bearing in mind that a plane like the 757 contains components contributed by some 50 major subcontractors, each at a different location, and such is the precision and strength required in a modern jet aircraft that if any two structural components fail to match by a margin of more than ten-thousandths of an inch, Boeing's rulebook requires that a metal shim be inserted to fill the gap. The first 747 required several hundred – the first 757 contained just six!

Even more remarkable, perhaps, was the speed with which that first plane was assembled. Based on their experience with the 747 and other major new projects, the Boeing engineers had scheduled over ten days for the assembly of the main fuselage structure and the wing spars of the first 757. In fact only two days were needed. Moreover, according to Beeby it only took that long because, when the quality control teams were summoned at the end of the first day and told the job was ready for inspection, they simply refused to believe that a plane which had gone together so fast and easily could be 'right', and insisted that the entire structure be dismantled and rebuilt on day two!

Overall, Boeing reckoned that the use of CAD saved something like a third of the man-hours which would have been spent on manufacturing the 757 had the design been done using traditional methods. The saving that time represents will obviously have made the additional costs involved in design work look trivial by comparison. Furthermore, the time that was saved at the manufacturing stage meant that design work did not have to be begun in earnest until later in the lengthy and complex process of conceiving and selling a new aircraft. Boeing was able to make use of that delay to keep its options open, adjust the specifications to meet customers' requirements and generally ensure that as many orders were in the bag as possible before metal was cut.

One final point about the case of the Boeing 757, a point which brings home the immense value of CAD not just as a mechanized drawing board but as an electronic testbed on which the behaviour of components, assemblies and entire systems can be simulated.

Testing a modern jet aircraft is an enormously lengthy and rigorous process. For example, if problems emerge and the number of failures in any particular area of the test programme exceeds a set level, then that phase of the programme will have to be restarted from scratch. In the case of the first wide-bodied jet, the 747, Beeby claims that 12 preproduction planes were built specifically for test purposes, an enormous investment in materials and manufacturing resources. When it came to the 757, Boeing estimated that their techniques had advanced to the point at which only five test aircraft would be needed. But it turned out that they had underestimated their progress: two of the five were not fully utilized. According to Beeby, this saving, too, was very largely due to the decision to rely on the use of CAD methods throughout the project.

The lesson which other businesses, even relatively modest ones, should draw from Boeing's experience is that it is vital to avoid the narrow and the short-term view when considering the pros and cons of a switch to CAD. In particular, it can be fatal to look at the issue only in the context of design as an isolated function unrelated to the rest of a business.

We have already suggested in Chapter 2 that design and manufacture are best thought of as a single continuum, and argued that the benefits of CAD will only be appreciated if it is seen as something which has an impact on that entire continuum, not just one part of it. Overall, undoubtedly, CAD has a tendency to enlarge the design function by transferring control over the manufacturing processes from the factory floor to the design office, and that tendency is greatly reinforced if CAM is added to CAD. In a very direct way, therefore, the introduction of CAD thrusts new duties and responsibilities upon the designer. Instead of simply allowing him to do what he did before, only faster or better, CAD makes it possible for him to assume a measure of control over the way in which a product is manufactured that was previously impractical.

Since CAD involves the designer in doing more work rather than less it is simplistic to expect that it will always lead to savings of time or money at the design stage. If we want to find savings then we must often look for them elsewhere: in manufacturing, testing, quality control, sales and even accountancy. They may not always be obvious or easy to quantify, but they will rarely be negligible either.

Saving Time – Availability of Information

On a rather different scale from the Boeing example, but of just as much relevance to the would-be user of CAD/CAM, was another potential CAD installation considered in the course of a consultancy assignment. The client in this case was a medium-sized British engineering firm involved mainly in the design and construction of specialized process plant, and the striking feature of the study was that two of the main applications which were identified for a CAD system were, strictly speaking, outside the design field altogether – certainly they would not have seemed relevant had CAD been considered simply as a tool for improving the productivity of the firm's designers.

In many cases, the company's involvement in a project began with an invitation to tender, often simply on the basis of a specification supplied by the customer, and, the process-plant business being highly competitive, it was estimated that only about one in twenty of the tenders submitted would actually lead to the signature of a contract. None the less, the preparation of block diagrams, showing the main elements of the plant needed to meet a particular specification, and the costing procedures necessary to produce a quotation based on those block diagrams, occupied a very high proportion of the time of some of the more senior design staff.

By creating special-purpose software which would allow any combination of standard components to be assembled as a CAD model, and by building costing and engineering information into the system, it was estimated that the time taken to produce a quotation might be cut back in quite spectacular fashion. Moreover, since the same system would also be used to design new components or to plan modifications or improvements to existing ones, it would be a comparatively simple matter to ensure that the information used in preparing quotations was kept abreast of all developments or changes in the company's range of products.

At the other end of the company's operations, so to speak, came the task of preparing instruction manuals and providing maintenance data. Since many of the installations involved were purpose built for a particular customer, and the customers were often overseas, this job was neither as routine or as trivial as it might sound. In many cases, an installation simply could not be assembled *in situ* without constant reference to detailed instructions, including exploded drawings, wiring diagrams, etc. Frequently, it was necessary for professional artists to be employed. They of course could not start creating, say, an exploded perspective drawing from the designer's drawings; they needed to have the actual assembly before their eyes so that they could refer to it. The timing of the whole operation from the trial assembly of a new plant in the factory through the creation of the documentation that had to accompany it to its final assembly, with the help of that documentation, on the other side of the world could thus become extremely critical.

But it quickly became apparent that virtually all these problems would vanish with the advent of CAD. The illustrative material required could be generated by the system, almost as a by-product, at very little cost and very short notice. This advantage, apparently incidental and certainly unlooked for at the beginning of the investigation, became quite a serious factor in the firm's decision to switch to CAD.

A different case study again showed how, by making geometrical information available easily and quickly, CAD may produce payoffs in unexpected places. The client on this occasion was engaged in producing high-precision mouldings of great complexity. Constant checking and a high standard of quality control were, therefore, vital factors in the business.

The existing inspection system called for periodic checks to be made on sample components collected throughout the production run. The inspectors were required to check for even the smallest defect and, if a fault was discovered, the run was halted and all components produced since the last successful check were destroyed.

Not only was the volume of production high, but the inspectors were also expected to cope with a wide variety of different components, each of which required a different set of checks and reference to the latest component drawings incorporating all modifications. Naturally enough, the inspectors felt themselves to be under constant pressure and their frequent queries created continual traffic, on paper and in person, between the inspection department and the drawing office.

It quickly became clear that CAD could do much to relieve this situation. If remote terminals were installed on the inspectors' workbenches and they were allowed direct access to the design information (though not, of course, the freedom to modify it) they would be able to answer 90 per cent of the queries that arose without either bothering the design staff or delaying their own work and, perhaps, holding up production while they waited for a point to be clarified.

It would also be perfectly possible, at the design stage, for all the information that the inspectors required to check a new component to be compiled. Thus, when production commenced an inspector would only have to identify the component to be provided, on screen, with everything he needed to know, including the readings he should look for on his gauges, the tolerances permitted, etc. This company, moreover, was far-sighted enough to recognize that it would be feasible to go well beyond this first stage. Ultimately it would be possible to build automatic inspection cells which would, upon identification of a component, not only perform geometric checks upon its dimensions by direct reference to CAD-generated data, but also verify the correct operation of assemblies of components.

It would be comparatively easy to provide further examples illustrating this theme – the case of the toolroom inspectors and their ball gauges related in Chapter 6 is another case in point. But we hope that enough has been said to suggest that the introduction of CAD into almost any business is likely, by providing a means of making geometrical information readily available when and where it is needed, to bring about savings in fields which may be far removed from the designer's traditional area of responsibility.

Saving Trouble – Analysis of Information

Analysis is almost certainly the most laborious and time-consuming aspect of design work; fortunately it is also the one which lends itself most readily to the use of computers. We have already seen that the calculation of physical or thermal stress by means of finite elements analysis is, so to speak, a 'natural' for CAD since it involves lengthy and repetitive number crunching.

But there it is also becoming clear that the use of kinematics and other forms of simulation may soon eliminate much of the need for experimentation with models and prototypes. Some aircraft designers, for example, are even beginning to talk of a time when wind tunnels will be superfluous since it will be feasible to simulate the behaviour of airflows and aerofoils mathematically.

Probably the best single example of a whole class of developments which may be in store is the use of software which simulates the behaviour of the materials used in injection moulding. The flow of material inside a mould – the sequence in which the various sections of mould and sprueways are filled, the precise size of the 'shot' required, etc – is, of course, absolutely crucial to the success of the process. Simply because plastic mouldings are produced in such vast quantities and are normally used in low-cost products, it is totally impractical to inspect or test any more than a small sample of any batch and the arcane skills of the mould designer are the first and by far the most important line of defence against faults.

But calculating the way in which a shot of molten plastic will flow through an intricate maze of channels such as, for example, those that form the 'trees' on which the components of the familiar model kits are hung, is a daunting task. If, as it might well turn out, various different arrangements have to be tried then the process will have to be repeated over and over again.

But given the proposed geometry of a mould, the necessary information about the raw material and the appropriate software, a CAD system can automate the whole tedious business and even present, on screen, graphic representations of the mouldings that will be produced under different conditions or with different sizes of shot.

This single example is sufficient, we would suggest, to indicate that any business in which substantial amounts of time and effort are devoted to analysing the properties of designs should look long and hard at what CAD may have to offer. Certainly, as we have already mentioned, companies involved in large-scale manufacturing, such as the major motor car makers, are already finding it worth while to invest heavily in the use of CAD not just for analysing the design of their products, but also for analysing the problems that will be encountered in automating their manufacture. It is a lot easier and cheaper to move a simulated robot around the screen in order to find the best position for it than it is to move a real robot around the factory floor.

Doing What Could Not be Done Before

Although most CAD systems will obviously have to be justified on the grounds that they will allow a business to do the things it is already doing better, faster or more economically and efficiently, it is important not to forget that CAD may also allow a business to do things that it could not do before. In some instances this can be because the skills required were simply not available prior to the advent of CAD – the example quoted on pages 145-146 is a case in point. In others it is because CAD offers an opportunity for looking at a design problem from a new angle and thus leads to new solutions being found for old problems.

A good example of this latter trend comes from the area of small mechanism design. Traditionally, the components of such things as camera wind-on mechanisms, for example, and the moving parts of many other familiar household gadgets, have been made of metal. The high precision required has made them costly to manufacture, and assembly has been an expensive, labour intensive process. The advent of CAD, however, has encouraged designers to explore the use of plastic mouldings, sometimes including integral hinges, as substitutes. In many cases the results have been spectacularly successful. Initially, the companies concerned had no skill in designing injection moulds and no knowledge of flexible mechanisms and it is certainly fair to argue that, were it not for their CAD systems, they would probably have never been able to make such a dramatic switch from old methods to new.

Chapter 8
Identifying the Needs of a Company

If there is one adage that applies with full force to the business of investing in CAD/CAM, it is: 'Look before you leap.' Look not merely at the range of equipment that is available, but at every aspect of the business which might be affected by its introduction. The point is not just that the purchase of a CAD system represents an investment decision on a rather more substantial scale than, say, the buying of new typewriters or filing cabinets, but that it also involves changes in the way that one part of the business, the design department, works, which may have ramifications, offer opportunities and provide benefits in areas which are apparently remote from the drawing office. Unless all these implications are considered right at the outset the chances are that problems will arise and opportunities will be missed.

The first step to be taken by any firm considering the purchase of a CAD system should, therefore, be a thorough and wide-ranging feasibility study. Properly conducted, such an investigation will be far more than a formality. But the emphasis must fall on the word 'properly': it is fatally easy for such a study to be too superficial, too narrow or too prejudiced. The person responsible for conducting it *must* be provided with the proper brief and the energy and authority to tackle it wholeheartedly.

Who Should Conduct the Feasibility Study?

Here, as elsewhere in this book, we are addressing ourselves primarily to the medium-sized to small business. If a giant multinational wants to investigate its own CAD needs it will, no doubt, have little difficulty in hiring the necessary expertise or finding it within its own ranks. For the company operating on a more modest scale, the matter is more problematical. Is this an issue like, say, the choice of a new car where the 'consumer' is in the best position to determine his own needs? Or is this a field in which the expert is indispensable?

There are, of course, experts aplenty, all willing and eager to investigate your needs and recommend how they should be met – at a price. The one price no business should be prepared to pay under any circumstances is the one which involves making a prior commitment to any particular brand of hardware. In order to avoid this it is necessary not only to spurn the advisory services offered by the manufacturers themselves, it is also necessary to treat many self-styled consultants with considerable caution — their independence from the manufacturers or their agents is in some cases based on little more than a technicality. Becoming a real expert on CAD and its applications should involve rather more than the acquisition of a chic address, an elegant letterhead and a vocabulary of jargon.

There are, of course, many genuinely independent consultants who are admirably qualified to advise on the subject. The problem, in many cases, is that they do not have the time, or you do not have the cash to spare, for the kind of in-depth investigation which a proper feasibility study demands. A consultant who, however worthy his or her motives, merely takes a superficial look at a company and then recommends some standard 'recipe' for applying CAD to its problems is doing no one a service. This is a field where virtually every business has needs and offers opportunities which are *sui generis*.

When a company is a subsidiary of another, or is in some other fashion linked to a larger grouping, there can be a strong temptation to assume that the necessary source of expertise will be found somewhere higher up the corporate structure. It may well be that the hierarchy contains an individual called the 'group CAD manager' or something similar, and it is perfectly possible that he is well qualified to answer questions on the subject; but that is not what is required. If it is to serve any useful purpose then the feasibility study must be conducted by someone who knows about CAD and asks questions about the business, not by those who run the business asking questions about CAD.

We stress this point only because we recently came across a firm which had fallen into just this trap of assuming that all would be well provided expertise was 'on tap' at group level, even though the individual concerned had neither the time nor the inclination to consider the project in detail. The result was a catastrophe in the course of which the subsidiary acquired an expensive CAD system without having any clear idea of how it could be used and without having made any of the necessary preparations, in terms of staff training, for its arrival.

Obviously, if a management asks itself the question: 'Who is best qualified to understand our business and determine if, how, when and where a CAD system could be of use to it?', the answer that frequently comes back is: 'Us.'

There are three dangers inherent in accepting this solution and making the feasibility study an internal affair. First, and most obvious, is the point that no member of the management team is likely to have a detailed knowledge of CAD. This can, however, be remedied to some extent by study of the available literature, including the present book!

Second, no management, however great its determination to seek objectivity, can study its own affairs with complete disinterest. There will always be aspects of the game which would be more obvious to a spectator than they are to the players.

Third, since no management team can be expected to abandon its day-to-day responsibilities and devote its attention to the CAD feasibility study, the decision to handle the project internally means, in practice, that it will be delegated to one or more individuals, and this, in our experience, is the root of most of the problems that are likely to arise.

However committed the people concerned may be to doing a thorough and dispassionate job, it is almost inevitable that personal and departmental prejudices will intervene. The production director, for example, may well be able to see exactly how the use of CAD would enable the designers to provide his staff with better information faster, but may equally well have a completely blind spot when it comes to identifying the ways in which the workings of his own department should be changed in order to take full advantage of the improvement. The chief engineer may be able to spot a dozen opportunities for speeding up the work of his designer-draftsmen by equipping them with interactive terminals, but may recoil in horror from the idea that he should abandon his own drawing board in favour of a workstation, and reject out of hand the notion of installing a terminal on the factory floor where production staff could use it to 'spy' on his department's work. The company accountant may insist that the benefits of CAD should only rank as benefits if and when they can be quantified in terms of cash saved.

It will be as well, therefore, to establish some ground rules right at the start. First, no aspect of the company's affairs, no matter how far removed from the scope of a CAD system it might at first appear to be, should be eliminated from the study *a priori*. The individual responsible must have *carte blanche* to probe everything from shop-floor practice to the managing director's working schedule if he thinks it relevant.

Second, all concerned must agree that the possibility of using CAD to make more money is just as important as the opportunity to use it as a means of saving money – an effort must be made, in other words, to balance the negative cost-cutting aspects by emphasizing the positive ones. No one is likely to cooperate very enthusiastically in a study which seems solely to be aimed at their job; but they may change their attitude if they understand that it may make their job better, easier, more interesting or more rewarding.

Third, and perhaps most important of all, it must be made clear that the study will not be a hole-in-the-corner affair designed to rubber stamp a decision already arrived at but, on the contrary, that the final decision will be a public decision based upon information available to all affected by its implications and arrived at publicly. It should be stressed at the outset that if the company does opt for CAD/CAM then it will be a venture which everyone will be involved in and which will, as far as possible, be to the benefit of all and the detriment of none. This is especially important because the study will only achieve its maxiumum value if everyone answers questions fully, frankly and without reservations – something that is unlikely to happen if people do not understand the purpose of those questions or suspect that they are aimed at eliciting information which will be used to their ultimate disadvantage.

There is absolutely no doubt in our minds that it is infinitely preferable for the feasibility study to be conducted by an outside agency, or even an individual, whose 'neutrality' is apparent to all. But we also recognize that it may be very difficult to find outsiders who are willing, and competent, to perform the task. It will probably be worth at least investigating the possibility that help may be obtainable from the academic community or from official agencies of one kind or another. But it may well be that, in the last resort, a company will have to give the task to members of its own staff; what follows is therefore primarily intended to help those unfortunates who draw the short straw to do the job as well as possible.

Geometrical Information – the Vital Commodity

Suppose, then, that for whatever reason it has been decided to make the feasibility study an exercise in corporate introspection and to make one of the existing management team responsible for it. Where should the inquiry start? What sort of information should be collected? What sort of questions should be asked, and of whom? What, in other words, should the investigators try to find out in order that the company may make an informed and intelligent decision about the role which CAD/CAM can and should play in its affairs?

As we emphasized at the beginning of this book, the key factor in CAD/CAM, the thing which it is *about*, is geometrical information. It follows that a study designed to establish the feasibility and value of using CAD/CAM will have to concentrate on the creation, movement and use of geometrical information. The point can perhaps be clarified by taking the analogy of a less esoteric area of 'computerization' – accountancy.

When a business switches from manual accountancy systems to computerized ones it will, naturally, take care to ensure that the new system covers every contingency, that no channel through which cash flows in or out of the business and no internal ledger account has been overlooked. The need for such comprehensiveness is, in this case, obvious. If, for example, the computer knows about the money that comes in from customers with credit accounts and the money that goes out in wages, but is never told about the proceeds of cash sales from the trade counter or the money that trickles out as petty cash, then it will not be long before financial chaos sets in.

Financial controllers, accountants and auditors have the advantage of professional training and years of experience in the techniques of keeping track of movements of cash and credit both within a business and between that business and the outside world – such, indeed, is the main purpose of their professional lives. Unfortunately, no such principles or methodology have been developed for 'accounting' for the movements of engineering information, and yet, as we shall try to show, it is in many companies a commodity that is almost as vital and as omnipresent as information about pounds and pence or dollars and cents.

In order to 'put it on a computer' the origins, comings and goings and ultimate disposal of geometrical information must first be detected and documented. Although this might seem like a task which, in any well-run business, should present few problems, experience suggests that it is not nearly so easy as it sounds. The persistent inquirer who interrogates not just the obvious people and departments but also the apparently irrelevant, who taps into not only the obviously geometrical channels of communication but also into conduits ostensibly devoted to other matters, will be surprised at the frequency with which a smidgin of geometrical material comes to light.

In order to make sure that this crucial phase of the study is carried out as thoroughly as possible it is worth sitting down and listing all the inputs and the outputs into the organization concerned, be it an entire company or one department or division of a larger firm. Consider, for example, the inputs into an average medium-sized manufacturing firm. They arrive via a number of different channels: by post, by phone and telex, by truck and in the heads and briefcases of both staff and visitors.

What, it may be asked, have such observations got to do with the feasibility, or otherwise, of using CAD? The answer may well be: quite a lot! The truckload of components that arrives from a supplier or subcontractor, for example, may have to be assembled together with other components produced in the company's own works – if CAD is to be installed then what administrative machinery will be required to ensure that a geometrical description of subcontracted components is captured and made available to users of the system? What about the telex from, say, the company's agent in Singapore? – may it not contain notification of a change in specification or a request for a quotation based on a modification of an existing item? Who needs to have this information, how could it be put on to the system and by whom? And what of the incoming phone calls, the hastily passed messages, the scribbled notes? – all of which may contain information that will have to be put into a CAD system rapidly, accurately and efficiently if that system is to be utilized to the best advantage.

Exactly the same sort of considerations apply to the organization's output. How, for instance, is the engineer out in the field to know that the factory decided that this bit of process plant would work better if the height of the gantry was trimmed by six inches if no one remembered to modify the information in the CAD system and, hence, the assembly drawings to which the engineer is working? What could be more likely to lose customers than modifying a product but failing to incorporate the relevant changes in the maintenance manual because no one thought to alter the CAD file and notify the department responsible for producing the manuals?

This sort of 'audit' of inputs and outputs can sometimes reveal facts which surprise even those most familiar with a company's internal workings. Not so long ago, for example, such a study revealed that an entire facility, supposedly handling all the electronic aspects of a large company's business, was in fact little more than a post office – no one had realized that, as a consequence of a series of reorganizations, the plant no longer had a *raison d'être*.

Where Does the Information Originate?

The obvious, orthodox answer to this question is that the design department is the well-spring of all geometrical information. But although this is certainly true of the vast majority of information in the vast majority of businesses there are exceptions, and the way in which they are provided for or taken account of can be important.

A company doing subcontract work on a regular basis, for example, may well find that the bulk of the geometrical information about its products originates with its customers. It will, therefore, be a first priority to ensure that any CAD system that it installs is compatible with those used by its customers – if it is not the company may, at worst, lose those customers or, at best, waste much time and effort translating information from one form to another.

An even more obvious case is that of the company which already uses some NC equipment. The kind of postprogramming required by those tools will clearly be a major factor governing the choice of CAD system. No one would want a system which made the preparation of part programs more rather than less tedious a task. Yet, surprising though it may seem, it is not unknown for companies to end up with CAD and CAM equipment that is mutually incompatible. And even if such extreme folly is avoided, it is worth remembering that the traffic of geometrical information between design and manufacture does not flow only in one direction. The design department, to take just one example, will need to be kept constantly up to date on the range of tools that are available, their sizes and their characteristics if the manufacturing process is to run at full efficiency.

The need for a CAD system to be able to accept inputs of geometrical information from a variety of sources, some of them perhaps far from obvious, is one side of the coin. The other is represented by the opportunity that arises for information which was previously disparate in nature and stored in different ways in different places to be brought together under one roof, as it were. We have already seen, in the case of Boeing, how great are the advantages that may be gained at the manufacturing stage if all geometrical material is generated in a form which is self-consistent. Another example will serve to illustrate the point that using a single system to create and store the information can produce substantial payoffs elsewhere as well.

The case in point was a medium-sized engineering firm in southern England which specialized in a range of equipment designed to serve a fairly routine industrial function. The established practice, when the company received an enquiry, was for the details to be passed to a project engineer who determined how far existing designs or stock components could be utilized and what new design work was likely to be involved. A design engineer then considered the implications of the necessary redesign work and finally passed the file on to a production engineer who would calculate the cost of producing the new items. The whole dossier then went back to the sales department where a final estimate was prepared.

Obviously, this system involved a labyrinthine series of communications, each of which contained at least some geometrical data. The fact that once a CAD system was in place all the necessary information would either already exist inside the system, or could be created with its aid, suggested a radical restructuring of the arrangements for producing quotations.

Under the new dispensation an enquiry would go directly to a 'CAD engineer', who would establish, using a fairly elementary version of group technology, how far the specification could be met using existing components which, with the system's help, could immediately be itemized and costed. Any additional components could then be designed, using the system, and an estimate made of the cost of manufacturing them. Finally a quotation for the entire job could be prepared. By putting the CAD engineer at the hub of the system, with access to all existing information and in a position to coordinate the production of new information as and when necessary, the whole quoting procedure was streamlined and greatly speeded up.

This example illustrates how vitally important is the recognition that the full value of CAD will only be realized if everyone concerned in creating geometrical information accepts and makes use of the system. Another case study, this time involving the production of process plant, emphasizes the point. The feasibility study in this case revealed that although the designer-draftsmen in the company's drawing office produced detailed drawings for each individual component, the only complete record of the way in which these components were to be assembled for any particular installation was usually a rough sketch produced by the chief designer.

Should the chief designer, heaven forbid, have fallen under a bus the rest of his staff would have had to wrestle with what was, in effect, a gigantic metal jigsaw puzzle with the help of little more than a few notes on the back of an envelope. Clearly, if individual components were all modelled on a CAD system, then it would be very easy for the chief designer to 'assemble' the particular combination required for any job on screen and file the description of the assembly away in the system where, if anything untoward happened, it would be freely available and easily understood.

In the case concerned the chief designer rapidly saw the advantages that the new arrangements would bring and realized that they would save him much time and trouble in that he would not be obliged, as was currently the case, to make himself constantly available in case problems arose during final assembly. But it is easy to imagine circumstances in which a less intelligent or more hidebound individual could, by refusing to change his methods of work, have effectively undermined much of the value of a CAD system.

How is Information Stored, Communicated and Used?

Even in businesses where geometrical information is the very lifeblood of day-to-day existence, it may be stored and communicated in a remarkably haphazard fashion. In practical terms, the assorted improvisations and working practices which are built up over a number of years may work well enough – provided that no major change occurs to throw the whole arrangement out of kilter. But the introduction of a CAD system is just such a change and the feasibility study must be based upon the premise that it will involve rethinking existing practices from start to finish.

This is likely to require a far more probing and radical enquiry than might be imagined. It is clear enough, for example, that linking CAD to CAM will revolutionize the way in which traffic flows along the super-highway linking design to production; but what about the byways that have come into existence in order to link design to the quality control, sales or warehousing and dispatch departments? What about the slip roads via which data from suppliers, standards authorities and customers is fed into the system? What about the archives which contain the records that must be referred to by maintenance staff or which may form the basis for the introduction of group technology?

It must always be borne in mind that, whatever the formal, 'official' channels of communication that exist within a business, they will always be supplemented by a network of informal, often concealed, ones. It is absolutely essential that the feasibility study should discover and trace all of these private information networks, which is not always an easy task nor one that is welcomed by those concerned.

It could be, to take a hypothetical example, that there is a formal system which quality control inspectors in a toolroom are supposed to use whenever they discover a component that is oversize. But it could well be that, rather than be bothered with all the paperwork created by that system, the inspectors have become accustomed to simply walking across the factory floor and handing the offending part back to the toolmakers – who are probably friends in any case – in order that it can be rectified without fuss or bother.

This could very probably be a highly efficient and satisfactory procedure from the company's point of view – time, administrative effort and ill-temper are, very likely, saved all round. But it will be very damaging indeed if the CAD system, when installed, has been designed to fit in with the 'official' procedure, that is, what happens in theory, rather than the 'unofficial' procedure, which reflects what happens in practice. Nor, it should be added, will CAD gain many fans if it results in the inefficient, but official, procedure being enforced at the expense of the efficient but unofficial one!

It is often very tempting to assume that the more wide-ranging implications thrown up by the feasibility study can be ignored, at least to begin with. The departments on the fringes of the company's geometrical circulatory system may well feel that the design department should show what CAD can do for them, at the heart of the system, and only then start disturbing everyone else's life. This attitude can lead to disastrous results. We would certainly not advocate any sort of overnight coup by which CAD was instantaneously imposed upon all aspects of the business, but the feasibility study must assume that this will *ultimately* come about, and assess the needs accordingly.

If, for example, the study is limited to the drawing office and its immediate concerns, then what happens in three years' time? Once the CAD system is up and running it may be found that the estimators in the sales department need access to it, or that much time is being wasted producing hard copies for the people in quality control who, if they had their own terminal, could obtain it direct from the system. But if these sort of problems were not considered at the outset then it is more than likely that the capacity of the system that it has chosen, or the way in which it has been installed, make its extension unnecessarily expensive if not totally impossible.

A classic instance of the sort of problems that can arise is provided by a major British company in the area of high-technology engineering, largely for the aerospace business. It being a very large company, the design and production engineering departments were allowed to go their own way when it came to determining their CAD/CAM needs. Initially the designers acquired what was, for its day, some very fancy equipment indeed, especially after it had been modified to meet their needs. A little later, the production engineers opted for a DNC system which, again, was technologically right up to the minute. The fact that information generated on one system now has to be specially prepared, a lengthy and costly process, for input into the other, was considered less important than the need for both departments to get the system that was as nearly as possible ideal for their limited needs.

Any business which wants to avoid even a small-scale version of this kind of counterproductive absurdity, which runs counter to everything that CAD/CAM should be about, must accept the logic of the situation. CAD and CAM, and their progeny like FMS and CAQC, and near relations like computer graphics, are all concerned with the same basic thing – geometrical information. It makes no sense at all to consider the different kinds of geometrical information, the different means of communicating it, or the different uses to which it is put, in isolation.

The Place of CAD/CAM in the Information Structure

The pace of change will, no doubt, vary from industry to industry and from firm to firm, but there is every reason to believe that within the next ten years geometrical information in businesses of all kinds will have become as thoroughly computerized as financial information is today, perhaps even more so. Whatever label is given to the technology which is devised to handle this function – whether it is called 'computer-aided engineering', 'computer-integrated manufacture' or any other of the trendy tags which are currently being suggested – it will essentially consist of a development of what we call CAD/CAM. The companies that are likely to reap the greatest benefits from this coming revolution are, we believe, those who accept that CAD/CAM offers them the opportunity to integrate the whole business of creating, transmitting, storing and using geometrical information.

This does not mean that there must be centralization of authority or loss of independence and initiative; it simply means that all concerned must recognize that geometry is the common currency of their working existence and that everyone will benefit if they use the same banking system.

The analogy is in fact a fairly exact one. A CAD/CAM system does operate as a sort of bank in which one set of customers deposits geometrical information in the form of designs, and other customers borrow or draw out that information and use it for their own purposes.

At the moment the problem is that any business contains many separate 'geometry banks', some heavily in credit, others dangerously insolvent, all using different currencies, all pursuing their own policies. Bring them together into a single system and instantly matters improve – cheques clear overnight, there are no longer losses due to converting between different currencies, service is fast and bank charges are reduced. Everyone's interests prosper. That is what CAD/CAM must hope to achieve.

Setting Identifiable Goals

But enthusiasm is no substitute for practicality. A conscious effort must be made by those conducting the feasibility study to make sure that a proper balance is maintained between long-term possibilities and short-term goals. There is nothing wrong with looking up to the stars, provided that you also keep your feet on the ground – but this is not always as easy as it sounds.

The promise of the new technology stirs the imagination and it is all too easy to get carried away. If models can be designed and tested on computers, if other computers can automatically produce the parts, if robots can assemble them and other devices can test them, and if all this automated production can be combined with computerized stock control and accounting procedures, then it might seem that the era of total automation is just around the corner. It is a seductive vision. But it is also a totally misleading one. Such a state of affairs, even if it is ultimately achievable, or even desirable, is still far over the technological horizon. It will demand computing resources and techniques vastly more advanced than those that are currently available.

Yet the vision, even if a mere fantasy at this stage, does contain elements which are relevant to present circumstances. Total automation, if it ever comes about, will require a system that allows all the functional elements of an enterprise – design, production, assembly, testing, etc – to communicate intelligently so that information, instructions, components and materials are always available in the right form, in the right place and at the right time. Currently most of the actual work must either be performed by human beings or supervised by them, but CAD does allow the basic infrastructure, the means of storing and communicating the information, to be put in place against the day when machines may, perhaps, be capable of taking a greater share of the workload.

The key to the integration of any process, whether or not it involves the actual manufacture of goods, is the provision of a data base and the system of communications that will allow that data base to be used to the best possible advantage. Whether the users are people or machines is, in a sense, a secondary consideration.

Thus, as the feasibility study progresses and a firm proposal begins to take shape, care must be taken to distinguish between those factors that are realistic in the context of the medium- or long-term future and those which are relevant actually to implementing a CAD system in the immediate future.

A couple of practical examples may help to clarify the distinction. The introduction of CAD into an engineering business, say, will allow a computerized data base containing full geometrical descriptions of all products to be created. The design staff working on the system will provide the raw material and it may be that the introduction of CAM will allow some of that data to be accessed and utilized in the course of manufacture.

But it is also highly likely that it will not be practical to provide many of those who rely upon data originating in the design department with direct access to the system; they will have to continue to make use of traditional drawings. A purist might argue that acceptance of this involves an intolerable degree of compromise. But, in fact, nothing serious will be lost in the long term provided that all the information gets into the data base. What would indeed be fatal would be to conclude that, because one particular category of information is not currently required to be accessible via the system, it should continue to be prepared manually.

Another example concerns a case study already quoted, where a firm producing process plant had, traditionally, separated the function of preparing the schematic drawings required for the purpose of producing estimates from the detailed design work that followed if and when a contract was actually captured. In this instance it was immediately obvious that everything was to be gained by allowing those responsible for both tasks access to a common data base. The time taken to prepare an estimate would be drastically reduced with the result, research suggested, of a dramatically improved capture rate. And the designers would have a head start when they began their work because the schematics upon which the estimate was based would incorporate geometrical models already in the system. In fact, it soon became apparent that the impact of a CAD system on the estimating procedures would, if anything, be even more significant than its advantages for the design process proper.

For a final example we could take the automation of inspection procedures. Again, this is a step that might well seem too ambitious or too costly to be considered as part of the feasibility study. But a failure to foresee a time when it might become practical could have disastrous consequences and when that point came involve an enormous extra cost. For it would be a comparatively simple matter to ensure that the necessary information was incorporated in the data base in the course of the design process, even if it was unlikely to be required in the immediate future. But the time and trouble that would be involved in adding it retrospectively could be prohibitive.

The essential point is that the feasibility study must set realistic goals at two levels. If the final proposal is too all-embracing it will not only be more costly than it need be, it will also arouse more internal hostility in that it will be seen as involving wholesale change and disruption throughout the entire business. The chances of it gaining acceptance will be correspondingly diminished on both counts. But if it fails to take account of long-term possibilities for future developments then it will, in the long term, prove to have been equally unrealistic.

Chapter 9
Choosing a System and Persuading the Company to Buy It

A CAD system can, as we have seen, consist of anything from a single microcomputer equipped with some special-purpose software and costing, perhaps, no more than a couple of thousand pounds in all, to a multi-million-pound installation. The main purpose of the feasibility study, assuming that it confirms that there is a need and a role for CAD in the business, is to narrow down this immense field of choice and to define what the proposed system will be used for, who will use it, and what will be required of it.

The feasibility study should, in other words, result in a pretty clear definition of what is wanted. This must then be matched against a list of what is available, constantly bearing in mind constraints such as budgets and the physical limitations of space, etc.

'Turnkey' Systems

The first choice that must be made involves either opting for one of the packages marketed by the major manufacturers in the form of a so-called 'turnkey' system or choosing to build a system up out of a combination of off-the-shelf hardware and software, plus, perhaps, some software devised in-house or commissioned from specialist suppliers.

In the vast majority of cases, there is no doubt that the balance of advantage is likely to lie with one of the turnkey systems. These will normally consist of complete workstations and all the requisite computer hardware together with a suite of standard graphics programs plus, maybe, a selection of programs directed towards specific applications.

In many cases there will also be a range of options available where hardware is concerned. The customer may, for example, be able to choose from half a dozen different plotters, all compatible with the system, or even from a number of different computers, rather as the purchaser of a motor car may be offered a choice of engines. This reflects the fact that most of the manufacturers of CAD systems are 'manufacturers' primarily of software rather than hardware.

Indeed, the distinction between turnkey and assembled systems is by no means so clear-cut, in many cases, as might at first appear. For whereas it is true that there are a few major manufacturers who actually produce not only their own workstations and display equipment but also their own CPUs, there are many other smaller fry who concentrate entirely on software and buy in all their hardware components from third-party suppliers. In some sense, therefore, what these smaller, more specialized firms are selling are often systems that they have 'assembled' on the customer's behalf.

It should be added, however, that the fact that the software developed by the companies which rely on outside suppliers of hardware will often run on a number of different computers – that is, that it is 'hardware independent' to a greater or lesser degree – may be very advantageous from the customer's point of view. Not only does he have a wider range of options when first purchasing a system, he may also enjoy much more flexibility when the time comes to develop it and, most important of all perhaps, if the manufacturer survives and keeps the software up to date it could well continue in use long after the current generation of hardware has been replaced by bigger and faster machines.

But, whatever the nature of a turnkey system, it does have a number of advantages over one which the user assembles for himself. The most obvious of these are that it can be delivered more or less ready for use; that some sort of training programme is likely to form part of the package; that there will, in theory at least, be a minimum of debugging to be done and what there is will, in any case, be the responsibility of the manufacturer. Similarly, if problems do arise once the system is up and running, expert help should be available on-call since virtually all the suppliers offer a service contract as part of the turnkey package. Third, the prospective purchaser of a turnkey installation should have little difficulty in finding out exactly what he is buying, what it can and cannot do and how much it will cost. Not only will the manufacturer or supplier provide the basic information, there are also likely to be established users, or user groups, who will be willing to share the results of their experience.

The disadvantages boil down, in the main, to a lack of flexibility both now and in the future. It is generally neither easy nor desirable to modify or add equipment from other sources to the package as delivered – manufacturers do not encourage users to 'improve' their wares and may well refuse to sort out the problems that arise when people try to do so. If, therefore, there is a real need for customized hardware or, more probably, software, it may be difficult to find a turnkey system that fills the bill.

Manufacturers also have a tendency to produce new systems at fairly regular intervals, just as motor manufacturers launch new models, and for very much the same reason. It subjects the customer to pressure, or if pressure is too strong a word certainly to temptation, to trade in a system (or at least software) that is now officially 'out of date' in favour of one that is brand-spanking new. And if, at the same time, just enough continuity is maintained between the old and the new to make a switch to a different manufacturer disadvantageous, then 'brand loyalty' is more or less assured.

The fact that the purchase of a turnkey system has the effect of binding all but the most determined customer hand and foot to a single manufacturer is a real problem, but one that is almost unavoidable in present circumstances. This makes it all the more important, from the customer's point of view, that the initial choice of system should be the right one. Again, the moral is: look before you leap.

Assembled Systems

There can, however, be little doubt that nine out of ten newcomers to CAD, the great central heartland of the market as it were, will continue to opt, and rightly so, for the known quantity of a turnkey system. The brave few who elect to try to assemble their own system will, for the most part, be found in the margins at the top and bottom of this market.

At the top are the multinational corporations and the giants of hi-tech industry whose CAD needs are either so large scale, or so esoteric, or both, that a purpose-built system may be the only one to make sense. Fortunately for them, their budgets are built to match, and they can both afford and justify the purchase of hardware and the hiring of programmers to provide the software.

More relevant to most readers of this book will be the opportunities which exist at the other extreme, where there may be a demonstrable case for the use of CAD but where it may be difficult to justify the cost of even the most modest turnkey system. In our view this may well be one of the major growth areas over the next ten years, and one which will break the existing mould in the CAD marketplace.

The pattern that may emerge could be rather similar to that which already operates in the case of computer systems intended for general-purpose business use. Here the running is made by the hardware manufacturers, most of whose products will run a wide range of business programs produced by specialist software houses. The sort of hardware we are thinking of is typified by the IBM PC, the Apricot range and the vast range of PC-compatible micros; while applications programs such as Wordstar, Lotus Symphony, Supercalc, etc, are familiar examples of the software.

General-purpose graphics programming is already easily available for most micros on the business market, and this is being rapidly extended to cater specifically for the CAD market. You can, for instance, buy CAD applications programs for the IBM PC and compatible machines, Apple markets a graphics tablet and interface for use with the Apple II, and other examples are easy to find. At present, this sort of development is having its principal impact in education and training – in the Brunel Special Engineering Programme, for example, students first encounter CAD via Acorn's BBC micro, a home computer *par excellence*, running software devised within the department.

But as CAD becomes a more familiar and less esoteric and intimidating field, more and more companies will realize that they, too, can 'do' CAD with the help of their existing microcomputers and some applications software which may be either purchased off the shelf or specially written. CAD at this level will not necessarily be seen as 'high technology' – any more than word processing or the use of a spreadsheet program is seen as high technology in an up-to-date office – but it may have more to offer the average, modestly sized business than some of the specialized gadgetry that ornaments the catalogues of today's CAD manufacturers.

The point we want to make is that the feasibility study may indicate that CAD could be useful at quite a mundane and humdrum level, and that such needs may be met by assembled systems built around the sort of hardware that is already in everyday use in many an office. If a company's CAD requirements are simple, and many are simpler than some 'experts' would have us believe, then they should be met by simple means. In our experience, companies select systems that are overly elaborate and powerful (and expensive) for their needs at least as often as they choose systems which lack the necessary power or refinements.

Sources of Information

Where do you look for objective, accurate and pertinent information about the systems and equipment that are on the market? There is, as the would-be customer will quickly discover if his interest becomes known, a wealth of glossy promotional literature. Aside from the pictures, which usually show the hardware carefully posed in pristine offices and manned by well-groomed individuals who look as if they would be more at home on the cat-walk than at the drawing board, the information contained in this material falls into two distinct categories – what might be called 'waffle' and 'jargon'. The 'waffle' basically says that all CAD, and this system in particular, is great, just great. The 'jargon' is intelligible only to people with a master's degree in electronic engineering.

In an attempt to compile some sort of catalogue of existing systems, which would provide a basis for comparison and a data base that a would-be user could refer to in order to see what a particular system can and cannot do, what its characteristics are, what the range of options offered consists of, etc, the staff of the Brunel Special Engineering Programme devised a questionnaire which is the basis of the checklist reproduced in Appendix II. Drawing on our own experience as users of CAD, and on the consultancy work which we regularly do for industry, we tried to concentrate on those features which were really significant to the user — not always quite the same thing as the features that the manufacturers are most anxious to bring to the customer's attention.

In the absence of any single source of up-to-date and comprehensive information, the reader who is contemplating the purchase of a system may find that the checklist provides a basic structure into which the information provided by manufacturers can be slotted as it is accumulated. Clearly, not every potential purchaser is going to be concerned with every point covered in the list; but it may serve as an *aide memoire* that will help to ensure that no significant points are overlooked when gathering data and that, when rival systems are being compared, the same set of standards has been applied to all. If nothing else, running through the questions on the checklist, or a selection of them which covers a customer's particular concerns, will force a salesman to come up with hard facts and abandon the routine patter.

There are, of course, plenty of consultants in the field, and we have already sounded a note of warning that not all are as independent or as knowledgeable as they might suggest; which is not to say that others are not admirably equipped to provide advice, providing, of course, that the customer is ready to foot the bill. Another source of information, less publicized and less frequently tapped, is the user groups which now exist in most countries and for most of the major brands of turnkey system. Manufacturers will usually be happy to put a potential customer in touch with his nearest group (certainly the customer should be suspicious of any reluctance to do so) and the information obtained in this way can be invaluable.

User groups will not only be aware of the major virtues and faults of a particular system as they have emerged in the course of practical experience, they will also be able to report on the manufacturer's past record. Was the training programme effective? Is the maintenance and emergency service up to scratch? Are customers regularly offered updates and improvements to existing software or are they confronted every five years by the 'need' to buy a new system? Are their complaints and/or suggestions listened to and acted upon? Talking to experienced users may also give the newcomer to CAD the chance to pick up a good many tips and a mass of informal knowledge that would otherwise be extremely difficult to come by.

The Politics of CAD

Although the above heading might suggest that what follows will be something of a digression, experience has shown that politics — the relationships between individuals and departments within a company and the way they fit into its organizational structure – are a crucial factor when choosing a CAD system. This applies especially in large, or largish, companies where whoever has been responsible for the feasibility study, it is not likely to have been the same people who will ultimately be responsible for authorizing the purchase. There will, therefore, be a need to 'sell' the project to the board, the executive committee, or simply to the boss.

The task will almost inevitably be complicated by the fact that the feasibility study, if it has been thoroughly conducted, will involve proposals that affect the way in which a number of departments work and, therefore, impinge upon the vested interests of a number of senior managers. We shall have more to say in the next chapter about 'selling' the idea of CAD to those who will be using the system; for present purposes it is enough to note that there will almost certainly be those who see it as being in their interests, or the interests of their departments, to resist the change.

There is always a danger that the team responsible for the feasibility study will 'sell' the idea to themselves so thoroughly that they too take up entrenched positions and become committed and over-fervent advocates of their own proposals. This tendency should be resisted, partly because too fierce a commitment in favour of CAD will almost certainly provoke equally determined resistance, and partly because it is vital that, whatever the final decision, it should not leave the company divided into two camps. If battle lines are allowed to form and a schism develops then, if the decision goes in favour of CAD, the business of introducing the system will be needlessly complicated and, if the decision goes against CAD on this occasion, it will be all the more difficult to revive the possibility at a later date.

If those who support the introduction of CAD are to make their case and win the day then they must go about the task with care, consideration and subtlety. Not only must they win over as much support and enthusiasm as possible at every level, they must also seek to disarm and outflank resistance rather than confronting it head-on. They must also, needless to say, ensure that their case is well founded, carefully argued and clearly presented.

It is fatal for the advocates of CAD to put on the mantle of visionaries who have 'seen the future' and are impatient with anyone who questions whether it will work. Talk of automatic factories, fifth-generation computers and a utopian future in which computers will take care of everything is not only unjustified; much more importantly in the present context it is calculated to terrify those who have to make the decision for or against CAD. They do not want to be offered what they may well consider little more than science fiction fantasies; they want a proposal which is hard-headed, sets realistic goals and is put forward with well-reasoned enthusiasm not futuristic fervour.

Experience suggests that the case for CAD is most likely to find favour with the people who have to find the money if at least one of those individuals can be won over early on and persuaded to fill the role of sponsor or 'champion'. Certainly almost every case study we have made of a successful CAD installation suggests that a vital role was played by some 'champion' figure who, because he himself was so firmly convinced of the arguments in favour, was better able to convince his colleagues. A reluctant champion is, of course, worse than useless, and so is a champion whose opinions do not carry sufficient weight with his colleagues at the top table.

If and when a champion emerges, or is recruited, then he must become the 'front man' for the campaign; for he will not only have the ear of his colleagues but he will, if he is the right man, also have the authority and influence to overcome departmental resistance and to put through the necessary organizational changes that will be needed if the system is to work properly. It is the champion who must tiptoe through the minefield of company politics and who must, ultimately, decide on the selling strategy to be adopted, and it is the function of those who support him to make sure that he is supplied with the ammunition he needs, that he is well briefed on technical aspects, that he understands the recommendations of the feasibility study in depth, and that he has grasped the case for CAD in all its aspects as well as understanding the strengths and weaknesses of the case against it. In the last resort, again it is the champion who must decide what can be achieved and where, if anywhere, compromises are going to have to be made.

The single most important question that will have to be decided by the champion, in consultation with his supporters, is which system should be recommended and on what grounds. That may sound silly – surely, it will be said, the choice should fall on the best system and the grounds for choosing it will be that it meets all (or as many as possible) of the needs identified by the feasibility study. However, as we shall now try to show, things are rarely as simple as that.

Making a Shortlist

No one, needless to say, is going to find the perfect system which meets their every need and has no snags at all. But the customer must ensure that he does not end up with a system that is unable to cope with tasks that, from his point of view, are central to its role in the business or which has drawbacks that could have been identified and avoided.

The selection process proper will almost always start with the drawing up of a shortlist containing, perhaps, four to six systems which seem to meet the criteria that will have been established by the feasibility study. Providing that the necessary information has been amassed beforehand, constructing a shortlist should be a relatively straightforward business.

We say 'should', because experience suggests that people often make the job a good deal more difficult than it need be. The vital thing is to determine the features which are, in any particular instance, vital and therefore non-negotiable. Normally these will cover basic characteristics such as the system's modelling ability, the number of terminals required (both now and, if expansion is envisaged, in the foreseeable future), the amount of memory, compatibility with existing equipment, if any, and the availability of the specific applications programs which are required.

If the feasibility study has been thorough, then the customer will have a very clear idea of his requirements on each of these counts, and he should resolutely resist any notion of compromise where they are concerned. Whatever its other attractions, for example, a system that has insufficient memory capacity to store all the programming and data that will be required on a day-to-day basis will be worse than useless – it will be a constant source of delay and frustration for years to come.

An example of how *not* to go about things was the suggestion (happily withdrawn) that arose in connection with a recent consultancy assignment. The committee charged with the preparation of the shortlist pondered all the factors that had to be considered and proposed that each system be awarded points on each count. The most suitable system would, it was argued, clearly be the one that notched up the highest number of points.

The folly of this approach becomes clear if we imagine applying it to the choice of, say, a new motor car. Suppose that the cars score ten points for each passenger seat and five points for each optional extra that is offered. It is entirely possible that on this basis we would end up choosing a two-seater sports car with electric windows, reclining seats and a host of other small luxuries in preference to a four-door saloon with a more spartan specification – which would be a fine and logical thing to do provided that we did not have a family to transport!

If a formula is required for the preparation of a shortlist, then the one we would favour is based on the use of Venn diagrams. Figure 79 shows the sort of thing we have in mind. Each circle includes all the systems that share one of the characteristics which, as a result of the feasibility study, have been determined to be essential. One, in this example, embraces all systems with basic drafting facilities, another all systems with the capacity for handling at least electrical circuit design, another all the systems which offer automatic tool path generation as part of the standard software package. There are, of course, many other factors that may be equally, or even more, important in any particular instance.

In this example, however, the fourth circle represents the budgetary constraints that have been fixed and includes only systems which are priced below £x. The need for such a financial ceiling is obvious. The cost of the system must, after all, be related to the benefits it is expected to produce in general terms, even if those benefits are not always easy to

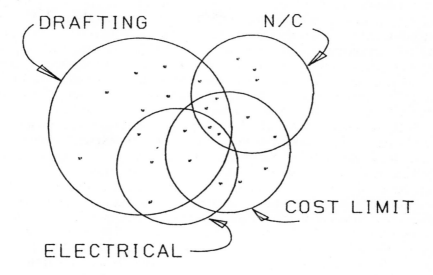

Figure 79. *The task of compiling a shortlist will be facilitated if a list of mandatory features, points on which the customer is not prepared to compromise under any circumstances, is first drawn up. The shortlist can then be narrowed down to include only those systems which meet all these basic requirements. The process can be represented, as here, in Venn diagram form. In this simplified example the mandatory features have, for clarity's sake, been limited to four: drafting, the capacity to prepare NC part programs, the availability of applications programs for electrical design and, finally, a price tag that is below a predetermined ceiling. The only systems, clearly, which qualify for inclusion in the shortlist are the two which fall within the area of intersection where all four circles overlap.*

quantify with precision. There may be other external factors which also have to be taken into account – the physical space available, for example, or the need to have the system up and running by a given date.

Using this technique, it can be seen that the shortlist virtually compiles itself. It consists of those two systems which fall within the area where all the circles overlap. Systems that fall outside this area of intersection can be dismissed from further consideration on the grounds that they will fail to meet one or more of the essential needs of the business.

It is worth stressing that the list of essential characteristics, the *sine qua nons*, must be compiled with intelligence and deliberation. If the selection process is to be tackled methodically and is to produce good results then the criteria must be matched to the real needs as determined by the feasibility study and, above all, they must be adhered to. This might seem self-evident, but experience has shown that it is a procedure more honoured in the breach than the observance.

In many cases people who start out with the very best intentions are tempted to stray from the straight and narrow path by the sheer glamour of the electronic 'goodies' on offer. It is fatally easy for the layman to be beguiled by the technology and to start thinking: 'Surely we could find a use for this spectacular kinematic software (or high-resolution display, or analytical programming, etc)?' This is a temptation that must be resisted. No matter what the charms of the equipment that the manufacturers flaunt before the customer, he must refuse to be seduced by them unless they are directly relevant to his needs.

Similarly, it is all too often the case that the effort and care that have been invested in a feasibility study and a selection exercise are negated because, at the very last moment, some 'irresistible' bargain turns up. The example that springs to mind is that of a CAD manager we recently met who described the elaborate procedure he and his colleagues had gone through in order to choose, as they believed, a system which matched their needs in every respect, only to have their recommendations overridden by the company's board of directors who suddenly discovered an alternative which failed to meet many of the needs but had the apparent virtue of being considerably cheaper! Like those who marry in haste, one imagines, that particular board may well find itself repenting its impulse buy at leisure over the next few years!

Benchmarking

So far, the selection process will have been conducted almost entirely 'on paper', and systems will have been included in the shortlist largely on the basis of their published specifications with, where feasible, additional input coming from established users or user groups. But once the choice has been narrowed down to a shortlist the time will have come to test the performance of the equipment in practice, doing the sort of work and tackling the sort of problems that it would actually have to cope with in the customer's own business.

The normal way of going about preparing a set of benchmark tests is for the customer to specify a series of tasks which, as far as possible, are representative of the work that is currently being done by those who will have to use the system if and when it is installed. An engineering company, for example, might choose a couple of the more elaborate piece-parts recently designed in its drawing office and ask for the working drawings and, perhaps, the NC part programs to be recreated on each of the competing systems.

This process will certainly establish that a system can do the work that is currently being done manually, and it will also show how quickly and easily it can be done on that system by comparison to others on the shortlist. But to rely on these standard tactics alone is, it seems to us, to miss an important opportunity. Suppose that, instead of the chief designer or engineer being asked merely for a sample that fairly represents what their staff *can* do at present, they are also asked for samples of things that they or their staff would like to do but currently find difficult or impossible.

An illustration of this approach is provided by a recent benchmarking exercise which was aimed at selecting a system for use in a large design department where much of the work centred on the design of complex, high-precision moulds. An essential part of this process was the calculation of the precise size of the cavities formed within the moulds and this, in turn, depended upon establishing the equations for the internal surfaces, a task which, using traditional drawing-office techniques, was immensely laborious and could never attain completely satisfactory levels of accuracy. When a benchmark test was needed, therefore, the chief designer was asked for the trickiest problem of this kind which he and his colleagues had encountered over the past couple of years – the one that they remembered with real bitterness – and that became the benchmark!

This approach to benchmarking has several advantages. It is likely, first of all, to provide a more rigorous and, therefore, more revealing test of the system's abilities. It may also help to identify applications for a CAD system which had not previously been considered. But arguably the most important point of all is that (providing CAD does come up with solutions) it will encourage those who face the prospect of having to use the system to look upon the change in their working lives positively and enthusiastically. It will make, in other words, a real contribution to the process of 'selling' CAD within the ranks of the company's own staff.

But it should not be thought that any set of benchmark tests, no matter how thorough and ingenious, will produce results that are totally clear and unambiguous. They will not 'prove' one system to be better than all others, but will, instead, show the relative strengths and weaknesses of all the contenders. In some cases, of course, a weakness may be so serious that it will eliminate the system concerned forthwith, but it is likely that at the end of the day several competing systems will still remain in contention; some will have turned out to be marginally faster or have better displays than others, while advantages such as convenience, a better menu or more memory space may redress the balance.

The 'Best' System?

How, then, can a final choice be made? Should it be left to a toss of a coin? Or determined, straightforwardly, by choosing the system with the lowest price tag? The problem is that the final shortlist will never consist of systems which all have identical specifications and capabilities; on the contrary, each will have its own particular pros and cons. How can they be balanced one against the other? Again a diagrammatic representation of the problem can help to clarify it.

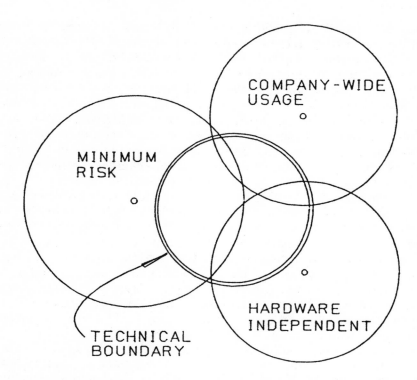

Figure 80. *Clarifying the choices. The 'technical boundary' in the centre includes all the shortlisted systems. In order to differentiate them and to distinguish clearly the available choices three 'poles' have been established outside the circle, each of which will 'attract' those systems sharing one particular characteristic – in this case one pole attracts systems that can be said to involve a minimum degree of risk on the part of a purchaser; the second pole attracts systems which, potentially, could form the basis of a company-wide system; the third pole attracts systems which can be described as 'hardware independent', that is, they consist of software packages which will run on several existing varieties of hardware and, more importantly, can be expected to run on the next generation of hardware as well.*

At the centre of Figure 80 is a circle which represents the 'technical boundary' containing all the systems on the shortlist, that is, those which were found in an area of intersection equivalent to that in Figure 79, less any that have been eliminated as a result of the benchmarking process. Initially, the shortlisted systems could be thought of as being scattered at random within the circle. But a number of 'poles' are then inserted around the perimeter of the circle, each representing one particular 'pro' enjoyed by one or more of the systems, and these poles 'attract' the systems causing them to form clusters in different parts of the circle.

Thus, in the example we have chosen, the first pole has been labelled 'minimum risk' and it will attract tried and tested systems produced by large, well-established manufacturers with, it is safe to assume, a minimum of bugs and a virtual guarantee that they will do everything that is claimed for them. The second pole, labelled 'company-wide usage', will attract those systems that could, in the future, be adopted by other divisions within the company, thus opening up the possibility of the entire organization standardizing on one system. Finally, a third pole, labelled 'hardware independence' will attract the one shortlisted system which is essentially a software package that can run on several different sets of hardware. Clearly, a system of this kind is likely to have a longer life in that, providing it is regularly updated, it can continue in use even when the hardware becomes obsolescent and has to be replaced.

Looking at a diagram like this – and it is, of course, possible to think of many other factors that could serve as the focus for a 'pole' – the customer can immediately see the choices which face him. In this case, for example, the systems fall into four categories:

1. There are two systems which offer minimum risk but neither the possibility of company-wide usage nor hardware independence.
2. There is one system offering company-wide usage but it does not afford either minimum risk or hardware independence.
3. Another system offers both minimum risk and company-wide usage but not hardware independence.
4. Finally, there is the single system which offers hardware independence but it is not suitable for company-wide usage and could not be said to offer minimum risk.

Clearly, options 1 and 2 can be eliminated; they both have less to offer than option 3. But the choice between option 3 and option 4 will ultimately depend upon which factors the purchaser considers most important and relevant to his particular circumstances. This, in turn, may depend upon a multitude of factors ranging from the purely technical to the entirely political.

It may be, for example, that keeping open the opportunity for the implementing of a single CAD standard throughout the entire company will weigh so heavily with those who must take the final decision to fund a purchase that it will prove overwhelming. But it is also perfectly possible to envisage circumstances in which the choice will fall on the hardware-independent system, the only one which, all other things being equal, may provide this part of the company with the foundation stone of a CAD structure that will have a working life of 30 years or more. Equally likely again, however, is the possibility that the dangers of choosing a system which may contain bugs and, therefore, take some time to 'run in' will seem too great.

Political considerations, on the other hand, may determine that a particular system should be recommended because it is more likely than any of the others to win the support of those who have not yet committed

themselves for or against the purchase of a CAD system – because, perhaps, a competitor had recently installed a similar system or, conversely,
because it is more advanced than a competitor's existing system.

Although the need to take political factors like this into consideration
may offend purists with a coldly logical attitude to the issue, they are
none the less all too prevalent, and nothing will be gained by ignoring
them. Here is the point at which the advice of a champion, the man as it
were who has to take the case for CAD into court and win a favourable
verdict, is likely to be crucial. He, after all, is in the best position to decide
what is achievable and what 'plea bargaining', if any, will have to be
undertaken.

Ready, Get Set . . .

The period before a final decision to purchase a CAD system is taken can
be a tense one for the team who have prepared the case. It is easy to
assume that, with all the work involved in the feasibility study done, the
choice of system to be recommended already made and the champion
fully armed for battle, nothing can be done but to sit back and wait.

This is not recommended. At the very least, support should be actively
fostered in order to retain enthusiasm and keep up the project's momen-
tum. Often the superficial inactivity will conceal a frenzied undercurrent
of lobbying within the company, with all concerned seeking information
and clarification as well, perhaps, as support for their particular point of
view. Meetings will be summoned, committees, official and unofficial,
will mushroom into existence, memos will fly hither and thither. It is likely
to be very counterproductive if the team responsible for the proposal try
to control or orchestrate this activity. But they should be aware of it and
must recognize that much of it arises because people feel unsure and
threatened; much can therefore be done by way of providing reassurance
or simply information which will help set minds at ease and lay some of
the wilder rumours before they do too much damage.

Within the team itself there may also be impatience and discord. Some
will want to press for a rapid decision and will become irritable in the face
of continued requests for further information or interrogation by the
decision makers. It is, however, fatal for the team to lose its nerve at this
point. Too much pressure for a decision is more likely to lead to a 'no'
than a 'yes' answer. Impatience with apparently 'stupid' queries or ignorance on the part of those responsible for making the decision will merely
turn them against the proposal. A positive attitude, on the other hand,
one which welcomes questions and willingly provides the answers efficiently, will build up the confidence of the decision makers and demon-
strate to them that the proposal is well founded and that those who support it have done their homework and 'covered all the bases'.

Whilst the team await an answer to their proposal, they usually attempt to lay down some plans for possible implementation. In our experience this is the wrong approach for a number of reasons. It puts them in a cleft stick because if they are seen to be producing the plans they are charged with 'jumping the gun', whilst if they do not they can be accused of being unprepared to act on a positive response if one is quickly given. The main objection to this approach, however, is that the proposal is incomplete if it does not contain a plan of the installations and introductory activities (a subject covered in the next chapter). Without such a plan no realistic cost justification and payback forecast can be generated.

The attitude adopted by the team should be one of being relaxed and ready. All should be aware that plans have been made to cover all possible outcomes. When the decision comes, whatever it is, the appropriate plan will be plucked from the shelf and implemented in a professional manner. No undue display of emotion will sway the decision makers. The team must show its professionalism and accept the decision. If it is to reject the proposal then such an attitude will allow a new proposal to be generated (built upon the previous work) which can be submitted at a later and perhaps more propitious time. A display of emotion as a response will only result in damaging the chances of another proposal succeeding in the foreseeable future.

If, however, the time and conditions are right and the case is well prepared then the team can look forward to a successful conclusion to their study. At such times we have been known to allow our professionalism to drop ever so slightly! After months of intense effort and nail-biting negotiations we would not be human if we did not feel pleased with ourselves and organize some small celebration. When it occurs it is a great feeling, to be fully enjoyed before commencing the real, hard work of purchasing, installing and commissioning the chosen system.

Chapter 10
Buying and Installing a System

Other than the obvious points of getting the most suitable system and the best possible value for money, there are a number of things which should constantly be borne in mind when making a deal for the purchase of the system which has finally been chosen. The customer must always remember that although the manufacturer may look upon the sale as just one of several currently in progress, the buyer is not only making a major investment, he is also taking a step which may radically change the way in which whole areas of his business operate. He is, or should be, concerned not only to make the 'best buy', as it were, but also to make sure that the whole process of installing the system, training the staff and generally switching from one way of working to another should be as orderly, untroubled and positive an experience as possible.

This means thorough and diligent planning which, if it is to be effective, will require the goodwill and cooperation of the supplier. It is not merely that the actual business of physically installing the system needs to be planned well in advance; training of staff (which will normally be part of the package if a turnkey system is involved) should take place as it best suits the buyer, not according to some inflexible schedule laid down by the supplier. We will have more to say shortly about the vital importance of organizing the training of staff. For the moment, the point to remember is that buying a CAD system is not like buying, say, some new office furniture. In making the purchase the customer is not only committing himself to the investment of a given sum of money, he is also committing himself and a substantial proportion of his staff and colleagues to an experiment which will only work if all concerned do their level best to make it work. This means that the supplier of the system must, in turn, be committed not just to making the sale, but to making sure that the system works to his customer's advantage to the fullest possible extent.

Beware, then, of the supplier who is anxious only to get a signature on the contract and has no real interest in what happens after that. Look, rather, for the one who makes it clear that he wants a happy and profitable relationship with a company over the long term and try, as far as possible, to make sure that the deal that is eventually struck reflects that commitment.

Implementation: the Role of the CAD Manager

The preparatory phase that follows a firm decision actually to purchase a system and precedes the arrival of the hardware on site may last several months and, if all is to go smoothly, every moment will have to be put to good use. The physical layout of the system must be planned and the necessary preparations made for its installation; staff must be trained, procedures set up, and a thousand and one decisions made on points ranging from the highly significant to the totally trivial. If all this preparatory work is to stand any chance at all of getting done there must be one person who takes responsibility for it – a CAD manager.

All other things being equal, the CAD manager may well be the person who headed the feasibility study or, if not the leader of the team, at least a member of it. Whoever he is, he will need to be determined, skilful at handling both his subordinates and his superiors and, above all, energetic. There will be an immense amount to be done: everyone whose work will impinge upon or be affected by the new system must be briefed and if necessary retrained, new channels of communication may have to be established, and new interdepartmental procedures agreed. None of this will happen of its own accord and, unless the CAD manager is there to drive the project forward from day to day, the chances are that the entire venture will get bogged down in committees or entangled in interdepartmental rivalries.

Even in the most auspicious circumstances, where everyone concerned is prepared to welcome the introduction of CAD with open arms, things can easily be overlooked or go awry simply because most of the people involved will be concerned with only those aspects of the planning process which directly affect them. If there is not one individual with overall responsibility for the project as a whole it is fatally easy for people to retire behind their departmental defences and make their own plans for using the system – plans which, it may subsequently emerge, are utterly incompatible with the plans being made by other departments.

This must be prevented at all costs. Much of the point of the system may be lost if, for example, the drawing-office staff sit down by themselves and plan the structure of the geometrical data base in order to suit their own requirements and without reference to the possible consequences for the work of other departments. It is the CAD manager's job to forestall disasters like this by making sure that no decision is taken without first checking and double-checking all the possible ramifications.

It is, therefore, essential that the CAD manager's role be recognized by everyone involved. This does not mean that his word should be law in all that concerns the CAD system. But it does mean that no one, however senior, should take any decisions involving the system, its implementation or its future utilization without first consulting the CAD manager and taking full account of what he has to say. He may not be the commanding general for 'operation CAD', but he is certainly the chief of staff and, as such, must be kept fully informed of everything that is going on as well as taking full responsibility for ensuring that everyone knows what is planned and what role they are expected to play.

Planning the Installation: Physical Factors

In a perfect world, of course, every new CAD system would be installed in a purpose-built setting. But, since industry and commerce must operate in a world that is far from perfect, most first-time purchasers of a system will face the problem of where to fit it into an existing physical plant and how best to arrange the various elements in order to make sure that they function effectively.

Figure 81 shows a couple of fairly standard workstation configurations. Note especially a point which is often concealed by the sort of glossy photographs that appear in manufacturers' brochures — the fact that the actual 'drawing' is done on-screen does not eliminate the need for desk space and storage space. Not only do the various items of hardware such as display screens, graphics tablet, keyboard and keypad, and hard copy unit have to be accommodated (and each designer must be given sufficient space to arrange these in his own idiosyncratic fashion if he is to work well), but there will also be a constant need to refer to the often bulky operating manuals and, very likely, to plans and other paperwork, all of which will have to be stored within easy reach. Nothing, in other words, is more likely to get the whole CAD system off to a bad start with those who have to use it than an effort to cram too many workstations, with too little workspace on each, into too small an area. CAD may be part of the new technology, but it must be so arranged that it can coexist with more traditional features of office life like coffee cups and personal filing systems that consist of a pile of paper with a heavy object on top.

If the choice has fallen on a system of self-contained workstations with no host computer, then both the station's own processor and its local memory facility can probably be accommodated within the space indicated in Figure 81. Remember, though, that even the best technology does go wrong and if the workstations have been crammed into too small an area then a small malfunction in one may result in the work of the entire department being brought to a halt while engineers shift the furniture around to get at the back of a display or trace a suspect cable.

If the system is to incorporate a service or host minicomputer then additional space will be required — probably in a separate room with air conditioning. Figure 82 shows the space occupied by a typical CPU, a disc drive and the systems console, which will always form part of such an installation. In every case, again, access space is crucial. There may

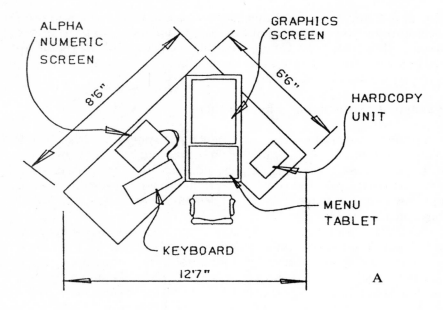

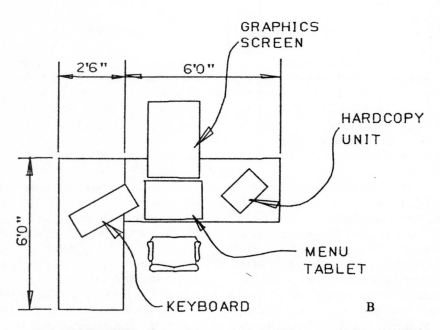

Figure 81 (A and B). *Two representative layouts for a workstation, with dimensions marked. Note that where several workstations are grouped together it will probably be necessary to provide some 'communal' work surfaces and shelving for items such as manuals and reference material.*

also be a limitation on the distance between a host computer and the workstations which it serves, or between the individual workstations in a networked system. This is something you will want to determine at an early stage, for it can, in some instances, even be an important factor in selecting a system.

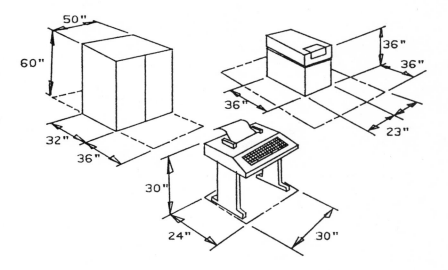

Figure 82. *The spaces occupied by (clockwise from the top) a typical CPU, disc drive unit and system's console. Where a system is built around a host minicomputer, or incorporates a mini to function as a service computer in a network, these three items will have to share a single air-conditioned space — indeed, there may well have to be provision for several disc drive units. In a networked system space will also have to be found for disc drives, which may either be incorporated in the workstations or installed as separate units.*

The actual positioning of the workstations themselves will also depend upon who is to use them and for what purpose. We will be considering the psychological and organizational repercussions in a moment, but there are also factors which are strictly a matter of physical layout. Where, for example, should the plotter be situated?

If there is a central computer with a 'room of its own', then that may also provide a home for the plotter. But remember that if a sheet-fed model is involved it will need regular attention. If, on the other hand, it is more convenient to have the plotter near the workstations, where its needs can be easily attended to, then bear in mind that some models can be quite noisy, sufficiently so as to constitute a source of perpetual irritation.

Figure 83 shows the space occupied by a fairly typical plotter (a flat-bed model would of course be quite a bit larger); also there will have to be provision for the storage of paper, spare pens, etc, plus, maybe, a convenient surface on which large plans can be laid out for examination.

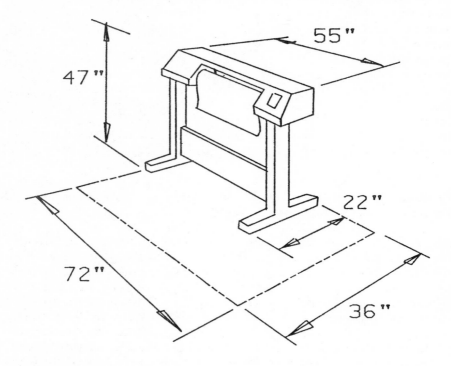

Figure 83. *This upright plotter is one of the smallest hard copy units available, other than the desk-top devices which supply nothing more than a relatively crude 'photocopy' of the screen image. A flat-bed plotter will obviously occupy a very much larger space. The siting of the plotter is an important consideration because of both the irritation caused by some machines and the need, in other cases, to replace the paper regularly.*

Whatever the layout adopted, lighting is going to be a crucial influence on the comfort and, therefore, the productivity of those who use the system. Nothing is more irritating in the short run and more likely, in the long run, to discourage the use of the equipment than display screens which are subject to glare or reflection. Also, the health issues raised by the need to peer for hours on end at an indistinct image are not negligible. The positioning of workstations must therefore take full account of the way in which a space is lighted, where the windows are and from which direction the sunlight comes.

A direct conflict of interests can also arise if CAD is to be used in conjunction with traditional drawing boards. For, as we shall explain shortly, psychological factors make it very counterproductive to isolate a

CAD system from the rest of the design department or drawing office; yet the lighting requirements of those using pencil and paper and those using CAD are utterly conflicting. One group needs bright, direct light; the other shuns it like the plague. Another physical factor to be borne in mind and thoroughly investigated at the planning stage is the power supply required by a system, which may range from a simple wall socket to a 'clean' source at a specified voltage with back-up facilities. Health regulations are also relevant. There may also be stipulations about working conditions that have been built into local or national agreements between unions and employers. But at least as much attention must be given to factors which are less easily quantified but no less important; these relate to the way in which the system will be viewed by staff and the way in which it will be incorporated into the working of an organization.

Planning the Installation: Psychological and Organizational Factors

An installation that has been badly planned in physical terms, with inadequate access space, say, or bad lighting, will certainly function with something less than 100 per cent effectiveness; but in the long term a failure to take psychological and organizational factors into account when planning a system may be even more dangerous. Often the worst mistakes in this area are made from the best of motives and supported by reasoning which seems impeccable. In some cases, in fact, those responsible never realize that a mistake has been made or how heavily they are paying for it.

The commonest and most straightforward error is to isolate the CAD system physically from the rest of an office or plant by putting it, so to speak, behind closed doors. All too frequently this has the effect of turning the installation into a sort of hi-tech shrine and its staff into dedicated acolytes who are set apart from their colleagues. In such circumstances it is very difficult for CAD to become integrated into the day-to-day working routine; instead it remains mysterious, the preserve of a privileged set of initiates, and this can be fatal. The staff of the drawing office, for instance, may never see the system as a useful tool but as an intimidating machine to which they turn only in the last resort.

Such an arrangement also leads to a situation in which one group tends to see the system as 'its' property or plaything and to guard it zealously against the incursions of 'outsiders'. If, for example, a single part programmer has to come and 'ask permission' of the design department (or even feels himself to be in that position) before using 'their' system then it is unlikely that he will ever feel much enthusiasm for it.

Ideally, those planning the installation should try as far as possible to arrange matters so that the CAD system is, figuratively speaking, placed where the action is, where it will be a focus of activity and interest and not something to be admired from afar. Some research done by Boeing illustrates the sort of point we are getting at very clearly.

Much of the company's detailed design work is done by small teams working under a team leader, and the idea was that each such team should have a single CAD terminal which any member could use as and when appropriate and through which the team as a whole could tap into the central reservoir of design information. The question was, where should this new tool be sited in order to ensure that it served the best interests of the team as a whole?

Boeing did not adopt the obvious answer, which would have been to tuck the workstation away in a quiet corner where individuals could go and use it 'in peace'. Quite the contrary. For their research showed that many of the team leader's responsibilities were handled informally; he would simply move round the office looking over people's shoulders asking a question here, making a suggestion there, thus not only keeping each team member's work under supervision but also maintaining a grasp on the work of the team as a whole. If one member decided on a minor design modification, for example, then the leader would quickly spot the implications for the work being done by other members and alert them to the need to take account of the changes.

This informal system of supervision, Boeing realized, was at least as important to the smooth functioning of the team as the formal one. But clearly it would quickly collapse with the introduction of CAD unless some means could be found of allowing the team leader to 'keep an eye' on what was being done on the screen in the same way as he monitored more traditional working methods. Once the need was recognized the answer was apparent: put the CAD workstation right next to the team leader's desk. It would then be easy for the leader to keep unobtrusively in touch with the use that was being made of the system; at the same time, anyone using the system could quickly involve the leader in a problem without having to make a formal approach.

The crux of the message for those planning the installation of a new system is that everything possible should be done to make the new tools accessible, convenient and approachable. Ideally, for staff who have a legitimate need, using the CAD system should be no bigger a deal than picking up the telephone to make a call.

Once upon a time, of course, telephones were exotic instruments and a transatlantic call, say, was something to be carefully considered. Those days have long gone and people talk to other people in different continents as a matter of routine. The sooner everyone involved comes to think of CAD in the same way, that is, as a means of making their working life more productive and efficient which they use as a matter of course, the sooner it will be used to best advantage.

The way in which a system is planned goes a long way towards encouraging this attitude. A workstation should not be a sacred space, set apart from the rest of an office. It should be another desk, albeit one with some useful built-in extras, where staff feel free to eat their sandwiches or drink their coffee. The room in which the workstations are situated may, it is true, have to be lighted with care, but it should not necessarily have a hushed or reverent atmosphere; if designers work best by talking to each other then they should talk as freely at the terminal as at the drawing board.

In short, those who will use the system should not be encouraged to feel that they are being transported out of their familiar working environment into a wonderland of technological mysteries, but rather that some extremely useful hardware is being introduced into the place where they already work.

Selling CAD to the Users

The above attitude will be greatly encouraged if all those who are likely to use the system, whether on a day-to-day basis or more irregularly, have been involved in preparing the proposal for switching to CAD right from the start. It is, for example, almost inevitable that rumours will start to fly from day one of the feasibility study. And, human nature being what it is, most of those rumours will be gloomy if not downright apocalyptic. It is essential that they be nipped in the bud before three-quarters of the staff become convinced that the introduction of CAD will mark the end of their jobs or at least the end of office life as they have known it!

The only satisfactory remedy is total frankness. This is indeed essential for other reasons as well, for, as we have seen, the success of the feasibility study depends crucially upon the present workings of the company, both official and unofficial, being laid bare. It therefore requires total cooperation on the part of the staff. That cooperation will not be forthcoming if the whole exercise is cloaked in mystery. People must be told exactly what is being considered, and why it is being considered, and, as the study progresses they should be regularly briefed on the team's thinking and the role which is foreseen for them under any new dispensation. Nothing is lost by adopting such a policy; almost everything that could be gained by the introduction of CAD will be lost by a failure to do so.

Again, as the project becomes more clearly defined, every effort should be made to stress the positive advantages that will accrue, not just to the company's finances, but to the working conditions of the staff and the solution of the problems that they encounter. We have already suggested that the requirement for a benchmark test can be turned to advantage in this connection by taking the opportunity to show staff that CAD may be able to resolve problems which they currently find troublesome or perplexing. Other practical steps can be taken. It may, for instance, be feasible to arrange for employees to attend demonstrations or, even more usefully, to visit plants where CAD is already in use and where they can talk to people who are already using it to do work comparable to their own.

It is, however, equally important that hopes are not built up to unrealistic levels. It would, for example, be grossly irresponsible to pretend that it will be possible to acquire the ideal system that will meet everyone's needs. It must be explained that, when it comes to the point where a choice must be made between competing systems, someone is probably going to be disappointed – though there will be no harm, provided no lies are told, in looking forward to the time when the system can be expanded to meet a need that cannot be taken care of right away.

Nor should management look upon the discussion and debate that will inevitably arise as a regrettable necessity. Time and again experience shows that suggestions that will actually enhance the eventual value of the system will come from the most unlikely quarters. It is worth reflecting, moreover, how much more likely staff are to use the system if some of the input for the proposal has come from them.

It is vital for this whole process of persuasion that CAD does not come to be seen as the 'baby' of one department or one group of people within the workforce. There will, assuredly, be those who dislike the idea and seek to form factions to fight it, and this resistance must not be allowed to become institutionalized or to crystallize on a departmental basis. One of the good things about CAD (and CAM for that matter) is that it does genuinely have something to offer by way of greater ease, speed, convenience or comfort to all who use it. Why not make a point of preaching this gospel from the start rather than creating an atmosphere of suspicion and fear?

Training

Once upon a time there was a company (which had better remain nameless, though we can vouch for its existence) which decided to purchase a CAD system. The first mistake it made was to take this decision in secret without conducting any study to determine its needs or making any attempt to discuss the project with its staff. But worse was to follow. It would be convenient, the management decided in their wisdom, to take advantage of the three-week holiday period when the office was closed to install the new equipment. As a result, the entire design staff returned from their summer vacation to find their accommodation transformed. Gone were the desks, drawing boards and filing cabinets and in their place stood ranks of gleaming new CAD workstations.

This is probably the most perfect demonstration yet provided of how *not* to go about installing a system. Little imagination is needed to envisage the instant fury and resentment felt by the staff, nor the chaos that ensued when they attempted to master their new equipment. This is one story that did not end with everyone living happily ever after!

Although it is difficult to underrate the importance of fully preparing the physical environment for the arrival of a CAD system – having detailed plans drawn up, the space cleared, the power supply ready, the cables laid, and so on – there is absolutely no doubt that the importance of a comprehensive and carefully planned training programme is immeasurably greater. Physical deficiencies can be remedied rapidly at the cost of no more than inconvenience and aggravation; lack of training may have consequences that will continue to plague the company for years to come, for CAD provides no exception to the rule that bad habits, once acquired, may be almost impossible to eradicate.

If a turnkey system has been purchased then there will normally be a contractual provision for the manufacturer, or third parties appointed by the manufacturer, to provide training – usually quantified in terms of a set number of modules each consisting of, say, one week's training for one employee. There may be different types of module to suit different needs – one kind, for example, for draftsmen, another for NC part programmers and another for the systems manager. In general the idea is that a representative selection of the more senior staff should be given a concentrated course of formal training and that they should subsequently become responsible for training everyone else. In most cases the contents of these training packages will be standardized, but the quality can vary considerably. As we have already suggested, every effort should be made to discover how effective the training provided by a manufacturer is likely to be before plumping for a particular system.

But, however good the formal training is that the manufacturer offers, it will only serve as a foundation stone upon which the CAD manager must try to build. As we shall explain in a moment, the first six months or so during which the CAD system is in operation will represent something of a trial run, and during this period there must be an intensive effort to develop skill and familiarity with the system by further training. This need not, and in many cases definitely should not, consist of formal 'classroom' lessons.

Younger employees, it is true, may still be close enough to their school or college days to feel quite at home listening to a formal lecture from a 'teacher'. But for older people the experience will not only be unfamiliar and disconcerting, it will also be threatening to their own self-respect. It is more than likely that older members of staff will have more difficulty than their younger colleagues in adapting to the new equipment (apart from anything else they are unlikely to have spent as much time playing with computers and video games!) and they will not relish their problems being exposed in public, so to speak.

Indeed, experience suggests that training will be much more effective if the trainers do not look upon the staff as 'trainees' who must be taught how to use CAD but instead see themselves as people with one set of skills which may be useful to the staff – people with another set of skills. Thus, rather than sitting everyone down in a row and addressing them like schoolchildren, the trainer will probably get better results by working on a one-to-one basis, asking each individual to explain his work and the problems involved in it and then demonstrating how CAD can help to solve those problems. Throughout, the key to success lies in demystifying CAD and presenting it as a tool – a tool which can help to get the job done faster, better or more easily – encouraging those who may be intimidated by the system to experiment with it and discover for themselves what it can do, and setting their fears at rest.

There will, of course, be some, often the younger members of staff, who take to CAD like a duck to water, and others, often older, who find the change harder to cope with. Whatever happens the former should not be allowed to lord it over the latter. Often those who learn more slowly learn more thoroughly, but nothing is more likely to discourage a fifty-year-old person from making the effort that may be required to come to terms with new skills than having a twenty-year-old person looking impatiently over his shoulder.

It is also worth remembering that the need for training will not be limited to those who are to be immediately involved in the use of the CAD system. Some sort of introductory or familiarization course should be provided for everyone in the organization from the top executives to the unskilled workers on the shop-floor whose job will in any way be affected. It should be made plain to them that the system is not an expensive toy which has been purchased for the sole use of the drawing office but that it will form the heart of a data base that can potentially be of value to almost everyone on the payroll. However remote CAD may seem to be from the concerns of an individual employee, he should be encouraged to understand the basic principles of the system and to speak up if he sees any way in which it might help him to perform his work more efficiently or effectively.

The First Six Months

It would be nice if, once the hardware was in place, the switch to CAD could be accomplished overnight. Alas, experience suggests that too much emphasis on immediate, short-term results will almost always be detrimental in the long term. No matter how thoroughly the CAD manager has prepared the ground there will be unforeseen snags. Bugs may be found in the system itself; those using CAD for the first time will work slowly and tentatively and with frequent resort to the instruction manuals; even those who have had the benefit of formal training will find that there are many things they do not know how to do.

We would, therefore, frankly urge that the first six months be treated as a period for training, experimentation and familiarization, with the system being run in tandem with existing pre-CAD practices. This will admittedly place a strain on all concerned and may involve a good deal of extra work, but in the long run the time and money will have been well spent if some of the disasters which can result from over-hasty implementation are avoided.

There is also the point that, inevitably, some of the decisions that were taken prior to implementation will turn out to have been faulty. It is therefore no bad thing to test all procedures for an initial period, after which they will be reviewed and, if necessary, changed. Surprising oversights may emerge, sometimes due to trivial causes. It has, for instance, been known for a company to set up a system for numbering drawings that only allowed for the recording of 999 items – a silly error but one which, if not remedied early on, could lead to countless complications later.

There will be a tendency on the part of some to say that, after running the system for six months and with perhaps several hundred files now created, it is better to stick to existing arrangements, however unsatisfactory, rather than go through all the laborious work that will be needed to make retrospective changes. In our opinion this line of argument should always be resisted. It is indisputable that changing the system at this stage will be inconvenient and will involve extra work; but it is better to have a few days wasted at this point than perhaps hundreds of days wasted over the next 20 years due to the company being stuck with a system that is needlessly, and by then irremediably, inefficient.

The review should cover every aspect of the system and its use and everyone who makes use of it or the information it contains should be encouraged to contribute. This, it should be pointed out, is their chance to say what they find inconvenient or unsatisfactory about the system or to make suggestions for improvements. They have now had a chance to see it in action and to live with it on a day-to-day basis; if they are not happy about things they should speak up now or subsequently hold their peace.

A simple example taken from recent experience will illustrate the sort of problem that comes to light in the course of a review. In the case of one newly implemented system in an engineering works it was found that, although provision had been made for layering models so that different kinds of information could be stored on different layers, no standards had been established for the way in which information should be categorized, or which categories should be allotted to which layers. As a result much time was being wasted searching through all the layers of a model in order to find, say, the one that contained the tolerances or, conversely, one which had no dimensions and therefore provided a clear isometric image.

Once the problem had been identified it was not difficult to devise a remedy, but if minds had not been concentrated by the formal procedure of a review the company might have struggled on for the next ten years without anyone ever getting round to tackling the problem. By that point, of course, it would have been quite impractical to implement any remedy retrospectively, whereas it was perfectly possible, if tedious, to do so after a mere six months.

Another matter that should be reviewed at the end of the initial six months is the level of staff skill that has been developed and the possible need for further training. Even if everyone appears to be 'coping' with the new system that does not mean that they are necessarily making the best possible use of it. It must be appreciated that no matter how thorough the training programme may have been, only so much can be absorbed in the form of 'book learning' and while there may be some individuals who go on to improve their skills by exploring the system and its potential, there will be many other less venturesome souls who, having once learned how to go about a task, never feel impelled to try and see if there is a way in which it could be performed better or faster.

It is, for example, quite common to find draftsmen who are using their workstations exactly as they would use a drawing board and have scarcely begun to grasp the opportunities that it should have opened up for them. Another blind spot that is often encountered concerns the need to pause before starting to construct a model in order to consider what it is to be used for, since this may radically influence the way in which it should be built up. If, for example, the aim is to establish the area of a cross-section then the model may be put together in one way, whereas the need to create a finite element mesh or to generate a tool path might dictate a totally different approach. Yet, all too often, a single 'orthodox' procedure is followed with the result that the opportunity to simplify a task or short-circuiting a procedure is not appreciated and much time and effort are wasted.

There are also cases where problems arise because some point which seemed, at the time, to be relatively unimportant was glossed over during the initial training period and as a result its relevance had never been understood. A classic instance of this was provided by a company which found itself perplexed, some months after its CAD system had been installed, because it was simply unable to get the NC package to work. The operator would put up on the screen what appeared to be a perfectly clear profile of a piece-part and punch in all the correct instructions according to the manual, but no tool path would appear.

Eventually, with some outside help, the roots of the problem were tracked down. The draftsmen who produced the drawings for the piece-parts that were to be machined had all learnt one particular way of inserting new entities in the course of constructing a model. This method, although perfectly satisfactory for their purposes, had the consequence that the entities were not mathematically related to each other but, instead, the position of each was recorded in terms of a separate set of coordinates. The information containing those sets of coordinates then became corrupted in the course of various computations. The extent of the corruption was minute and would have been totally unimportant were it not for the fact that, as far as the computer was concerned, the model no longer had a continuous perimeter but consisted of a series of lines which did not *quite* meet each other. The gaps where the lines met – or rather did not meet – were far too small to be detectable on the display screen, but in the system's own terms their existence was perfectly 'real' and made nonsense of the instructions which it was being given – how could it create a continuous tool path to follow a profile that was riven by discontinuities?

The manufacturers had foreseen this difficulty and there was a single, perfectly simple command which would cause the system to unify the outline of a model if it contained such unintended gaps. But this procedure had somehow been glossed over during the training period and, although it did feature in the instruction manual, it was not easy to find unless you knew what you were looking for – which those struggling with the problem of the 'faulty' NC package did not – and why, after all, should they? They had no reason to suspect that none of the lines in their apparently impeccable drawings actually met at the corners!

We also stress the need to review the level of skill with which the system is being used because there is a definite tendency for the collective skills of any group of people involved in the use of CAD to reach a sort of plateau of mediocrity, if not actually to decline. The reasons for this are clear; in nearly every group there will be a majority who, once they have mastered the basic skills needed to 'drive' the system to their own satisfaction, have little or no interest in exploring its potential any further. In due course, therefore, what might be called the 'folklore' of the group – its collective knowledge of the system and how to use it – will crystallize at a level which may be competent but is unlikely to be innovative or adventurous.

Naturally enough, in the years to come, it will be this folklore that provides the basis for the training of newcomers. Matters will not improve because, as time passes, there is likely to be positive resistance to new ideas. An attitude of 'this is the way to do it because this is the way we have always done it' will grow up and, if not regularly challenged, will result in the system being sadly misused and underexploited.

This failure on the part of users to make the most of CAD is partly the responsibility of manufacturers who, as we have already suggested, tend to lay the emphasis on crude measures of productivity and rarely take the trouble to investigate how a system might best fit the often idiosyncratic needs of one particular customer. But ultimately those who invest in CAD (and CAM for that matter) have only themselves to blame if they do not make the best use of their system.

We would, therefore, conclude by stressing the need to approach the new technology with an open mind. The opportunities that it opens up are enormous and potentially of great value, but they will not be realized unless those who use it are willing to think radically. This does not mean that changes should be made wholesale; it does, however, mean that users of CAD must be willing to experiment piecemeal and at a very practical level.

For the immediate future, then, we believe that CAD is likely to bring the greatest benefits to those who refuse to be blinded by visions of an electronic wonderland in which everything will be automated and look, instead, for the improvements that can be made by taking full advantage of the existing technology.

In the last resort CAD and CAM are tools like any other. They will be most effective in the hands of those who take the trouble to understand the principles involved without becoming so absorbed by the technology that it becomes an end in itself. We hope that this book will make some contribution towards ensuring that those who use CAD/CAM see it as it should be seen – as an immensely helpful servant that should never, never be allowed to become a master.

Glossary of Terms and Acronyms Used in CAD/CAM

Algorithm. The general method or procedure adopted in order to enable a computer to perform a task or solve a problem. Every computer program will, of course, be based upon an algorithm. But whereas a program is one specific set of instructions set out in a form which can run on a computer, an algorithm is a much more general concept and any one algorithm can be translated into a program in many different ways, using different programming languages, etc. The term is widely used in all branches of computer science.

Applications Program(ming). A program or set of programs designed for some specific application. The most commonly used application programming in CAD is, of course, *graphics programming* but there are also many other varieties designed for more specific applications in fields such as analysis or *numerical control.* An applications program must be distinguished from those other programs such as the *operating system* and the display control which are concerned with the functioning of the system in general terms rather than its use for any specific application.

Archival storage. The storage of computer-generated data in some more or less permanent form, often on magnetic tape. Unlike information kept in *back-up storage*, archival material will not be accessible *on-line* to the system.

Automatic dimensioning. A feature of most CAD systems which allows the computer, drawing on the geometrical description of a *model* stored in the data base, to add the annotations and other information such as dimensions, tolerances, datum points, etc, which are needed if a particular view of that model is to be an accurate and reliable source of reference. The title is to some extent misleading in that the procedure is by no means totally 'automatic' (see page 121).

B-rep. See *Boundary representation.*

Back-up storage or memory. Information stored on back-up devices, usually *disc drives* of one sort or another, will be available to the system within a matter of milliseconds, but not instantaneously as is the case with information lodged in the computer's *core memory.* Also known as virtual memory since, for many purposes, this form of storage is virtually equivalent to core memory, as the information it contains is rapidly accessible and can be transferred in and out of core memory automatically.

Bit. One binary digit, that is, a 0 or a 1. The basic 'alphabet' of the *machine code* which is the fundamental language of any computer's internal operations.

Bit map. In a *raster display* system, where the screen image is made up of a regular array of picture points or *pixels*, the intensity and/or colour of each pixel will be encoded in a sequence of binary digits, or *bits* of information. The complete description of an entire image is therefore known as the bit map or pixel map and will normally be stored in the *buffer memory* associated with the display unit.

Boolean operations. The algebra devised by the nineteenth-century logician George Boole is, of course, the basis of all computational processes. In the context of CAD, Boolean operations refer to the processes used to combine the *primitive solids* used in *constructive solid geometry* (see page 100).

Boundary representation (B-rep). One of the two approaches used in *solid modelling* – the other being *constructive solid geometry*. In B-rep the geometry of a model is defined in terms of the edges and surfaces that form its boundaries; the mathematical techniques involved are thus a natural extension of those used in *wire-frame* and *surface modelling*.

Buffer memory. The specialized memory elements associated with a *raster display* in which the *bit-map* of the image currently appearing on the screen is stored. The use of a buffer memory frees memory capacity in the *graphics computer* itself since the computer only needs to intervene in the control of the display when the image is to be changed or amended or transformed. A buffer memory is also known as a framestore.

CAD. See *computer-aided design*.

CADAM. An alternative acronym to CAD/CAM, currently fashionable because it implies a greater degree of integration between CAD and CAM.

CAD/CAM. The combination of *computer-aided design* and *computer-aided manufacture* – the subject of this book!

CAE. See *computer-aided engineering*.

CAM. See *computer-aided manufacture*.

CIE, CIM. See *computer-integrated engineering or manufacture*.

CNC. See *computer numerical control*.

COM. Computer on microfilm.

CPU. See *central processing unit*.

CRT. See *cathode ray tube*.

CSG. See *constructive solid geometry*.

Cathode ray tube (CRT). The essential element in all display systems.

Central processing unit (CPU). The logical circuitry at the heart of a computer which actually 'does' the computations. A *microprocessor* is an entire CPU crammed onto one silicon chip. The term is often used more generally to describe the cabinet containing the CPU together with its associated control units, input and output equipment, etc, that is, the 'computer' itself as opposed to *peripherals* such as *disc drives*, terminals, etc.

Compiler. A program which translates a *high-level language*, such as will normally be used by a programmer, into *machine code*.

Computer-aided design (CAD). The technique whereby geometrical descriptions of 2-D or 3-D objects can be created and stored, in the form of mathematical *models*, in a computer system. Once created, the models may then be displayed in a variety of ways. In many cases the models may also be analysed or their manufacture or operation may be simulated by the use of various *applications programs*.

Computer-aided engineering (CAE). An alternative term for *computer-integrated engineering*.

Computer-aided manufacture (CAM). The use of computers to control, wholly or partly, manufacturing processes. In practice the term is normally applied only to computer-based developments of *numerical control* technology, that is, it is not used to refer to techniques such as the computer control of flow processes used in the chemical industry.

Computer-assisted part programming. The use of computers to help in the preparation of the instructions, or *part programs*, which contain the data used to control manufacturing processes by one or other of the methods of *numerical control*. Computer-assisted part programming will normally involve the use of geometrical data prepared in the course of the *computer-aided design* process.

Computer graphics. The use of computers to create graphic images. In theory the term covers everything from the use of graphic techniques to present, say, financial data to what is sometimes known as 'computer art' and, of course, it includes CAD. But it is normally used mainly to describe the more advanced applications such as those employed in flight simulators, for example, or to provide special effects in the cinema.

Computer-integrated engineering and manufacture (CIE, CIM). The extension of the techniques used in *computer-aided manufacture* to create larger systems in which a number of computers may be interconnected in order to form a hierarchy that can control and integrate a sequence of manufacturing operations.

Computer numerical control (CNC). It is differentiated from the alternative method of using computers to control machine tools, *direct numerical control*, by the fact that each tool has its own separate computer *tool control unit* which normally accepts only one *part program* at a time.

Constructive solid geometry (CSG). One of the techniques (see *boundary representation* for the alternative) used in *solid modelling*. CSG involves the use of *primitive solids* which are then combined by means of *Boolean operations* to form geometrical models of more complex forms. The mathematical techniques employed are quite distinct from those used to construct other kinds of model.

Core memory (or storage). The electronic elements which provide the computer with its instantaneously accessible *random access memory* in which will be lodged the program currently being run and the data that are immediately required. Although core capacity will normally be supplemented by *back-up storage* the amount available within a computer, especially a *micro*, is an important measure of the machine's abilities. Today core memory consists of silicon memory chips, but previous generations of computers relied upon cumbersome and extremely expensive magnetic core stores.

Cursor. The luminous symbol which the operator steers around the display screen in order to point at a position on the screen or to indicate an alphanumerical character or, in CAD applications, the *entity* to which an instruction refers.

Cutting plane. The imaginary surface along which a *model* is 'cut' in order to produce a cross-section.

DNC. See *direct numerical control*.

DVST. See *direct view storage tube*.

Declare, to. To inform the *graphics computer* of something that, although it may be obvious to the operator, would not otherwise be evident to the machine. It might, for example, be necessary to 'declare' that a given space represented a 'hole' or that the area enclosed by four lines represented a surface and, therefore, that any lines 'behind' that area should be treated as hidden.

Depth cueing. Techniques for making a complex image more intelligible by helping the viewer to distinguish between elements that are in the 'foreground' and those that are 'further away'. In the absence of a facility for *hidden line removal* this may be achieved by varying the intensity of the lines according to their 'depth' (see page 93).

Digitizer. Also known as a puck, a digitizer is a device looking rather like a small magnifying glass and incorporating a crossed-hair 'gunsight' which can be very precisely positioned on a *graphics tablet*. It provides the only means by which the geometrical information on a paper plan can be input into a CAD system with a reasonable degree of accuracy.

Direct beam refresh. One of the two *stroke-writing* display systems, the other being *direct view storage tube* (DVST). Like DVST, direct beam refresh uses a steerable *electron gun* which actually 'draws' the lines that appear on the display screen; but, unlike DVST, it requires the image to be constantly redrawn or refreshed. Problems arise when the screen becomes so crowded that it can no longer be refreshed frequently enough to avoid the appearance of flicker.

Direct numerical control (DNC). Unlike *computer numerical control*, a DNC system employs a single computer, usually a *minicomputer*, to control the operations of a number of machine tools simultaneously. The advantage lies largely in the fact that the memory capacity of the computer and its associated *back-up* devices allows a whole repertoire of *part programs* to be stored in a form that is immediately accessible. A DNC system is, therefore, the natural choice when several separate processes are to be integrated. See *computer-integrated manufacture* and *flexible manufacturing system*.

Direct view storage tube (DVST). One version of *stroke writing* in which the need to refresh the image constantly is eliminated by the use of a storage tube, a screen which continues to display an image more or less indefinitely once it has been 'drawn'. The main disadvantage of DVST as compared with other display systems is that the screen must be completely 'repainted' every time an *entity* is moved or erased, see *floodgun*.

Disc drive. A device which allows information to be both recorded on to and read off magnetic discs at very high speeds. Although there are many different varieties of discs and thus of drive units, the fundamental division is between *floppy discs*, which are relatively cheap and easily stored, and have a capacity measured in hundreds of kilobytes (a kilobyte contains 8000 *bits* of information), and *hard discs*, which are expensive and need to be kept in sealed storage units, but have a capacity measured in megabytes (a megabyte is eight million bits of information). Any sizeable CAD system will contain at least one hard disc drive.

Electron gun. The device that actually creates an image on the display screen by firing a stream of electrons at its inner surface thus causing the phosphors which coat it to glow.

Entity. One of the graphic elements, most commonly a line, arc, circle or surface which forms part of a *model*. See also *macro*.

FE (analysis, technique, package, etc). See *finite elements*.

FMS. See *flexible manufacturing system*.

Feed rate. In machining, the rate at which the tool moves along or across the workpiece.

Finite elements (FE). By dividing a large or complex object in some regular fashion into a large number of separate elements and then considering the forces acting on each element separately it is possible to calculate where and when the object may distort or break under physical or thermal stress. Because both the creation of the *meshes* which divide the elements from one another and the subsequent stress calculations are immensely long this is a technique that more or less demands the use of computer-based methods.

Flexible manufacturing system (FMS). An FMS will normally consist of a 'cell' containing perhaps two or three machine tools, maybe some gauging or calibration equipment and a robot that will be responsible for loading and unloading the tools, etc. An FMS represents a sort of half-way stage between *computer* or *direct numerical control* systems, which are concerned only with controlling the operations of the tools themselves, and *computer-integrated manufacture* which aims to control and coordinate long sequences of operations.

Flood gun. A special form of *electron gun* used in *direct view storage tube* display which floods the entire screen with electrons and, by ensuring that special persistent phosphors in the coating of the screen continue to glow once they have been activated, allows the image to be 'stored' on the screen.

Floppy disc. Used to provide *back-up storage*, generally in micro-based systems, 'floppies' as they are familiarly known, have a far smaller capacity than the larger and more expensive *hard discs*.

Framestore. See *buffer memory*.

Graphics computer. The computer actually responsible for handling the creation of geometrical models and their display, as distinct from other computers that may, for example, be involved in *numerical control*, etc.

Graphics pad. See *graphics tablet*.

Graphics programming. Software that is concerned with the creation of geometrical models, their display, transformation, etc, as distinct from other *applications programming* that may be designed to deal with analysis, *kinematics*, and so on.

Graphics tablet. A flat pad, not unlike a blotting pad, with a grid of wires running beneath the surface which allows the position of a stylus device or a *digitizer* to be registered with great precision. Part of the tablet consists of a blank space, corresponding to the area of the display screen. Pointing to a position on this space will cause a cursor to appear in the appropriate place on the screen. The remainder of the tablet is normally occupied by a menu so that the operator can convey routine instructions to the computer simply by pointing to the appropriate box in the menu area.

Hard copy. Sometimes used to mean any reproduction on paper of an image derived from a CAD system, by whatever means it has been generated, but also see *hard copy unit*.

Hard copy unit. A small device, usually kept on the work surface at the CAD operator's elbow, which will very rapidly produce a relatively crude copy, on paper, of the image currently appearing on the display screen. Most hard copy units can be used only in association with *raster display* and none is an adequate substitute for a proper *plotter*.

Hard disc. The *back-up storage* device most frequently employed in CAD; as well as the standard hard discs there are a number of smaller varieties. See *disc drive*.

Hardware independent. If software has been designed so that it will run on a variety of different computer systems then it is said to be hardware independent. The significance lies partly in the freedom which this affords the purchaser; but software which has this property is also likely to continue to be of use well into the future when existing hardware may have been replaced by faster and better machines.

Hidden line removal. The facility for erasing lines, especially in a *wireframe* model, which would be hidden were the 'surfaces' of the frame to be filled in. Some systems can perform this task automatically; in other cases it can only be done by identifying the lines and instructing the system to delete them – a procedure which is, of course, wasted when the *model* is *transformed* hiding some lines that were previously visible and causing others to come out of hiding!

High-level language. One of the familiar programming languages – BASIC, Pascal, COBOL, etc – which people normally use to communicate with a computer. The language in which the majority of CAD software is written is FORTRAN. It will usually be necessary for the computer to make use of a *compiler* to translate information from a high-level language, which is understood by the programmer or operator, into the *machine code*, which is a computer's own native language, so to speak.

Host computer. A central computer that controls a number of remote *terminals* or *workstations*, each of which has some computational resources of its own, normally a *microcomputer*. The host machine will serve as a central resource, tackling tasks that are beyond the capacity of the smaller computers and providing a library of programs and a data base upon which they can draw. A system based upon a host computer will normally be configured in a star shape, with the host at the centre and the workstations at the tips of the lines of communication that radiate from it. This is in contrast to *networks* which, if they do contain a single large machine, will utilize it as a *service computer*. A host will normally be a *minicomputer*.

Intelligent (workstation, plotter, etc). A unit that is wholly or partially controlled by its own computer, normally a *microprocessor*, thus providing it with a degree of intelligence and making it partly independent of larger computers elsewhere in the system.

Interactive (graphics, system, etc). A computer is said to be used interactively when the process involves a 'dialogue' between man and machine. Virtually all CAD systems are interactive.

Isometric view. A 3-D view of an object so constructed that all lines that would in reality be of equal lengths are equal – that is, the distortions of the true values due to perspective are ignored.

'Jaggies'. The result of the step-effect that is to some extent apparent with most *raster displays*. Sloping or curved lines will run diagonally across the regular pattern of *pixels* and can, therefore, only be approximated with the result that they may appear extremely jagged on systems with low resolution.

Keypad. A smaller version of the familiar keyboard, usually with only nine numerical keys, which are used as 'function keys' to signal, for example, the responses to a menu displayed on the screen. Some *digitizers* may also incorporate an even smaller version of the keypad.

Kinematics. A technique that allows a CAD system to simulate the movement of images representing, say, the various components of a mechanism or, perhaps, the operations of a robot arm. Kinematic programs vary from those that operate relatively slowly and are used chiefly for analytical purposes to the very advanced ones which can realistically simulate the movement of complex objects in *real time*.

LAN. Local area network, see *network*.

Language. See *compiler, high-level language, machine code*.

Layer. In its most literal sense the term is used to describe a 2-D *model* which represents a 'slice' through a 3-D object; when stacked one on top of another a series of such layers produces a *two-and-a-half-D model*. But it is also possible to talk of a 3-D model having layers, in which case each layer will contain a different kind of information, rather like a master plan and a series of overlays. In architecture, for example, one layer might be used to store detailed structural information, another the layout of the services such as water and electricity and yet another the floor plans. Since layers can be viewed separately or in any chosen combination this arrangement has very clear advantàges when dealing with plans for complex structures or mechanisms.

Light pen. Used in conjunction with specially designed display equipment, a light pen allows the operator to 'draw' on the screen or, more frequently, to point to positions on it or to *entities* already being displayed. It is also, in some cases, used to indicate the operator's response to a menu displayed on the screen. It is an alternative input system to the *graphics tablet* and the various *cursor*-steering devices.

Machine code. When a computer processes information it takes two sequences of binary digits, or *bits*, and combines them in one of a number of different ways in order to generate a third sequence. All information that is to be processed must, therefore, be reduced to this form, called *machine code*. See also *compiler* and *high-level language*.

Macro. A macro is created when several *entities* are combined to form a graphical unit which can be treated as if it were a single entity. Some macros may be built in to the software, so to speak, and the operator will normally be able to create others as and when required. When a macro is common to the entire system – that is, can be used by anyone using the system rather than by one user alone – it is known as a 'system macro'.

Mainframe. Until the late 1960s virtually all computers were mainframes which serviced tens or even hundreds of terminals and performed many disparate tasks. But the hegemony of the mainframe was destroyed with the arrival of, first, the *minicomputer* and, then, the *micro* and the increasing tendency to dedicate a computer to one particular task – for example, controlling a CAD system. Today, the borderline between minis and mainframes looks increasingly blurred as the smaller machines grow in size and ability and most CAD systems are either mini- or micro-based.

Memory. Technically, a computer's memory consists only of its *core memory* and its *back-up* capacity, or virtual memory, should be described as storage since it is not instantaneously accessible. In practice, however, memory and storage are often used almost interchangeably.

Mesh. In order to divide a *model* up into separate elements for *finite elements* analysis a mesh must be generated. A similar mesh, consisting of a lattice-work of lines connecting the nodes or control points whose positions have been calculated by the computer is also used to represent surfaces in *surface* – and some *solid – modellers.*

Micro (-computer, -processor). A microprocessor is an entire *central processing unit* incorporated in a single chip and will furnish a microcomputer with its CPU. Micro, used by itself, usually refers to a complete microcomputer. Micros are widely used in CAD applications. Today, for example, most workstations will contain their own micro and it is even possible to do much CAD work on off-the-shelf micros designed primarily as home computers or for personal use in business. *Computer numerical control* involves equipping each machine tool with its own micro-based control unit. Super-micros are machines which have grown to the point at which they may rival the power of the smaller minis.

Mini, Minicomputer. A computer intermediate in size and capacity between a *mainframe* and a *micro.* Where CAD systems incorporate a central *host computer* or a *service computer* it will almost always be a mini; minis also provide the computer element in a *direct numerical control* system.

Mirroring. Duplication of a *model* by instructing the system to generate a mirror image of it. A technique that can be extremely useful when dealing with elements, such as wheels for example, which are symmetrical in form.

Model. When a design is 'drawn' on a CAD system the computer records the position of each *entity* in terms of coordinates in the model space, the imaginary 2-D or 3-D space that 'contains' the image, and its size, orientation, etc. This mathematical description of the geometry of the design is known as the model and is quite distinct from the image actually being displayed. In fact, by referring to the data which define a model a CAD system will be able to display many different images, each representing one view or aspect of the model.

Model space. See *model.*

Mouse. One of several devices – others being the joysticks and the *tracker ball* – used to steer a *cursor* around the display screen. A mouse consists of a small hand-held block running on a ball which can rotate freely in all directions; when the mouse is moved over a surface the ball turns, causing the cursor to move in a corresponding fashion.

NC. See *numerical control.*

Network. A group of computers, in CAD normally *micros*, interconnected so as to form a ring-like structure, in order that information may be exchanged and common resources such as *back-up storage*, a *plotter* or the resources of a larger *service computer* may be shared. Many contemporary CAD systems consist of workstations linked together to form a network; the alternative structure is built around a central *host computer.*

Numerical control (NC). The automatic control of machine tools by the use of preprogrammed sets of instructions. Invented in the early 1950s, NC originally involved the use of *part programs* which were written manually and then punched on to paper tape, in which form they could be 'read' by the *tool control unit*. Today *computer-assisted part programming* is widely used and most tools will operate under *computer numerical control* or *direct numerical control*.

On-line. Information, whether programs or data, is said to be on-line when it is stored in a form which is immediately accessible to the computer – that is, in *back-up* rather than *archival storage*.

Operating system. The basic programming which administers the computer's own internal workings and its communications with *peripherals*, and allows it to understand the *applications programs* that are loaded into it.

Panning. It is possible to 'move' the notional centre of vision represented by the centre of the display screen across an image just as a film camera pans over a scene. Panning will often precede *zooming* in order that the right bit of the image is 'enlarged'.

Parametric modelling. Parametrics involves varying the size and/or orientation (ie, one or more of the parameters) of the entities in a *model* in order to create different shapes while not altering the basic geometry of the model. To take a very simple example, a square can be transformed into a rectangle by shortening or lengthening one of the two parallel pairs of sides.

Part program. A sequence of instructions used in *numerical control*, so called because the usual arrangement is to write one program to cover the sequence of operations that one tool must perform on one part or workpiece. See also *computer-assisted part programming*.

Peripherals. Units such as *disc drives*, *plotters*, etc, which are connected to a computer and either provide a resource upon which it can draw or are 'driven' by it; they will normally be under the control of the computer's own *operating system*.

Pixel. One of the picture points or dots which make up the image on a *raster display*.

Pixel-map. See *bit map*.

Plotter. A device which reproduces on paper graphical information stored in the CAD system. Unlike a *hard copy unit*, which can only reproduce the image as it actually appears on screen, a plotter has access to the data in computer memory and can, therefore, reproduce one or more views of a *model* with great precision.

Postprocessing. When a *part program* is first written it will be in one of a number of languages commonly used in *numerical control*. In order to make it intelligible to one particular tool it must be translated into a more specific form by means of a program called a postprocessor.

Primitive solids. The basic 3-D forms such as cube, cylinder, wedge, pyramid, etc. They form the basis of the *solid-modelling* technique known as *constructive solid geometry*.

Processor. Any electronic element which actually processes information. See *central processing unit*, *microprocessor*.

Puck. See *digitizer*.

Random access memory (RAM). Computer memory elements in which any address (ie, any 'pigeon hole' in the memory) can be accessed with equal ease and speed, and which can, therefore, be used to store information of every kind, either programs or data depending upon the task to be performed. Most of a computer's *core memory* will consist of random access elements.

Raster display. A display system, similar to that used in a household TV, in which the screen is, in effect, divided up into a regular grid of *pixels*. The resolution of the system is measured by the number of lines and the number of pixels each line contains. A small micro-based display system might use less than 500 lines, a high-resolution one as many as 4000. The alternative to raster display is one of the *stroke-writing* systems.

Real time. A computer is said to operate in real time when it responds to an operator without delay or when, for example, it performs a simulation as rapidly as the action that is being simulated would take place in 'real life'.

Repaint. To repaint the display screen is to 'wipe it clean' and construct a completely new image from scratch. Any *transformation* will involve a repaint. One of the drawbacks to *direct view storage tube* display systems is that it is necessary to repaint the screen every time an *entity* is moved or deleted.

Resolution. See *raster display*.

Ring. See *network*.

Rubber banding. A technique used in drafting whereby one end of a line is 'fixed' in position on the screen while the other remains 'attached' to the *cursor*, which can then be manoeuvred around the screen, with the line stretching but remaining always taut like a rubber band, until it has been positioned to the operator's satisfaction. A similar technique is sometimes employed when adjusting the arrangement of elements that are interconnected, like the various units on a circuit board. In that case rubber banding allows one element to be 'picked up' and repositioned without any of its links with other elements being broken.

Service computer. A larger machine, normally a *minicomputer*, incorporated into the ring structure of a *network* in order to provide greater computational power and memory capacity for use by the other units as and when required. Unlike a *host computer*, a service machine is not the centre of a system, just one unit in it.

Silicon.disc. Introduced comparatively recently, silicon discs are memory chips that are treated by the computer and its *operating system* as if they were *discs*. That is, they are used to provide *back-up storage* rather than as *core memory*. Their main advantage is the additional speed of access which they afford when compared with real discs.

Solid model. A 3-D *model* in which not just the edges and surfaces of an object but also its volume are mathematically represented. Currently there are two very different approaches to solid modelling used in CAD: *boundary representation* (B-rep) and *constructive solid geometry* (CSG).

Stand-alone (workstation, etc). A stand-alone workstation is one which contains its own built-in computer, *disc drives*, etc, as well as display and input equipment. It is, in other words, able to stand alone and operate entirely autonomously. But provision may also be made for connecting a number of stand-alone units together to form a *network*.

Storage. See *archival storage, back-up storage, memory.*

Storage tube. See *direct view storage tube.*

Stroke writing. The alternative to *raster display*, a stroke writing system employs a steerable *electron gun* that actually 'draws' lines on the display screen. There are two versions of stroke writing: *direct beam refresh* and *direct view storage tube* (DVST). Stroke writing is also sometimes known as vector graphics.

Surface model. A *model* in which both the edges of objects and the surfaces which connect them (but *not* the spaces contained by these boundaries) are mathematically represented. Surface modellers usually represent surfaces on the display screen by a *mesh* of lines linking the control points, the points on the surface whose position the computer has established by solving the surface equations for the model.

Tape reader. Where *numerical control part programs* are prepared in punched tape form a tape reader will be attached to the *tool control unit*. The term is also applied, in CAD rather than CAM, to the equipment which reads data off magnetic tape for input into the computer system.

Tool controller, control unit. The device which actually controls the movements of an *NC* machine tool. Early units relied upon electro-mechanical relays and instructions were fed in via a *tape reader* a few at a time. Today most control units used in *direct numerical control* and all those used in *computer numerical control* are based upon *microprocessors* and have the capacity to store an entire *part program*, thus eliminating the need for the tape to be 'read' over and over again.

Tool path. The actual course, as defined in an *NC part program*, which is to be followed by a cutting tool. In many CAD systems tool paths can be computer generated on the basis of the geometrical description of a workpiece already stored in the computer's memory.

Tracker ball. An input device which allows the operator to guide a *cursor* round the display screen by turning a ball which is free to rotate in its mounting in any direction. Essentially a tracker ball is a *mouse* the other way up!

Transformation. A *model* is said to undergo a transformation when the computer 'redraws' the image on the display screen in order to present a different view – for example, to rotate a model, substitute a perspective view for an orthogonal one, or to *zoom* in on one bit of the model.

Two-and-a-half-D model. When several 2-D *models*, each representing one *layer* or 'slice' through a 3-D object, are stacked one on top of another the result is sometimes called a 2½-D model – an everyday example is provided by contour lines on a map which provide a 2½-D model of the landscape.

VDU. See *visual display unit*.

Vector graphics. See *stroke writing*.

Virtual memory. See *back-up storage*.

Visual display unit. Another name for any *CRT* that is used to display computer-generated information.

Wire-frame model. Best described as a 'drawing in three dimensions', a wire-frame model can represent the edges and corners of objects but not their surfaces or volume.

Workstation. A CAD terminal including, at a minimum, a display screen, a keyboard and some means by which the operator can 'point at' positions or *entities* on the screen plus, of course, access to a *graphics computer*.

x axis. One of the three axes to define positions in the *model space*. The x axis is best thought of as a line running across the display screen, the y axis as a line running down the screen from top to bottom, and the z axis as a line running 'back into' the screen from its surface.

xy or xyz coordinates. The coordinates that are used in order to define a unique position in the *model space*; two coordinates will be required in a 2-D modeller, and three in a 3-D system.

y axis. See *x axis*.

z axis. See *x axis*.

Zooming. A *transformation* which generates a 'close up' of one area of a *model* rather as a zoom lens can be used to show a scene at varying degrees of magnification.

Appendix II:

Checklist for Potential Purchasers of CAD Systems

The questions in this checklist were originally formulated in order to produce a questionnaire which the staff of the Special Engineering Programme at Brunel University sent out to all suppliers of turnkey CAD systems in the United Kingdom; many of these are, of course, British subsidiaries of companies based in the United States or elsewhere. The questions, however, can be applied with equal validity to domestic or foreign-owned suppliers in the United States or, indeed, anywhere else in the world.

It can, of course, be safely assumed that basic technical data, information about current prices and other material will be readily available in the manufacturers' published literature. The questionnaire did not therefore seek this sort of easily accessible information. What it, and consequently this checklist, were designed to achieve was to elicit facts which may not be so easily ascertained but which are likely, in the long term, to be of vital importance to the prospective purchaser of a CAD system.

Naturally, not everyone will be interested in the answers to all the questions and these should be modified, eliminated or added to in order to cover specific applications and address points which are of particular concern to a business. We reproduce the list in full, however, in the hope that it will serve to remind would-be buyers of the range and variety of considerations that they should have in mind and thus allow them to compile their own questionnaire for submission to potential vendors.

1. Company Information

Please supply sets of company accounts for both the local operation and the parent company, and a brief history of the organization.
1.1. When was the company formed (i) in the UK; (ii) in the USA?
1.2. Who was the original owner?
1.3. What is the current ownership?
1.4. Where are the administrative and technical centres for system development?
1.5. What was the company's turnover for the last three years (i) in the UK; (ii) in the USA; (iii) world-wide?

211

1.6.   What has the level of investment in system development been in the last three years?

1.7.   What is the planned level of investment in development?

1.8.   What is the relationship between the UK/US operation and the main development centre(s)?

1.9.   Are any organizational changes either in the UK/USA or world-wide planned? If so what are they and how will customers benefit?

1.10. How many CAD/CAM systems have been installed by your company (i) in the UK/USA; (ii) in the rest of the world? State how the above numbers are derived (eg, number of CPUs, companies).

1.11. What is the current structure of the UK/US company? (eg, departments, formal responsibilities, organizational tree)

1.12. What is the current structure of the parent company, with particular reference to system development and relations with the UK/US organization?

2. Availability of Support

2.1.   What is the number of vendor personnel dedicated to support (not development of new products) (i) in the UK/USA; (ii) in the rest of the world, in (a) hardware and (b) software?

2.2.   What is the number of personnel in each software specialism in (i) the UK/USA; (ii) the rest of the world, in the following: (a) Graphics/modelling/drafting; (b) NC; (c) User language; (d) Developing user applications; (e) Operating system; (f) Communications; (g) Others (specify); (h) in total? (Personnel should not be counted twice; if their time is split it should be accounted for by applying fractions. Hence totals should be as in 1.)

2.3.   Where are the UK/US support activities centred for (i) hardware; (ii) software?

2.4.   If there are additional facilities state their location and function.

2.5.   Is all support and maintenance coordinated through a single contact person? Who is it or how is it coordinated?

2.6.   How are service and support personnel assigned to customers?

2.7.   What is available as regards immediate organized/formal phone consultation for software and hardware problems and queries?

2.8.   What arrangements can be made for the vendor to develop applications programs for individual users and at what cost?

MAINTENANCE CONTRACT

Please supply a copy of the terms of your maintenance contract.

2.9.   Where are bug fixes and patches written and how long on average does it take?

2.10. What is the guaranteed maximum time between problems being reported and having them fixed?

2.11. What commitments are undertaken by the company if immediate repair is not possible (eg, penalty payments, replacement kit)?

2.12. What warranty is given with the hardware and software?

2.13. Is all support and maintenance performed by the vendor?, and, if not, then by whom?

2.14. How often are new revisions and enhancements available, and is there a commitment to release them regularly? Are there any additional costs involved?

2.15. What preventative maintenance is performed on a regular basis?

2.16. What is your guaranteed up-time and mean-time between failure? How are these statistics calculated?

2.17. Is it possible to have a non-standard maintenance contract to suit particular needs and have the cost related to the level of service?

2.18. Is there a remote diagnostics facility? What form does it take?

2.19. What is your policy towards ensuring users do not have obsolescent systems?

3. Future Development Plans

3.1. What developments are currently being undertaken, and when are they planned to be finished?

3.2. Which of these developments will be available to the user via the software maintenance contract?

3.3. What costs are involved in the user acquiring developments not covered by the maintenance contract?

3.4. How many people are dedicated to new developments of the system (not support of existing software) (i) in the UK; (ii) in the USA?

3.5. What is the commitment of the company to IGES (set up to establish international, industry-wide standards in CAD)? What level is currently supported? Is the company involved in the IGES committee? What other standards are the company considering?

3.6. Is development for specific users undertaken (eg, interfaces to analysis packages)? What is the cost and what conditions are applied?

3.7. Where are system developments performed and to what level?

3.8. What is the current level of investment in development for the system being quoted?

3.9. Who determines what developments are undertaken?

3.10. Is the company committed to ensuring upward compatibility of the system?

3.11. To what extent are new hardware developments incorporated into the system development and does this compromise upward compatibility?

3.12. Which operating system does the system run under? If more than one, is it the intention eventually to standardize?

3.13. What influence does the user base have over the development programme, and how are their requirements indicated to the development team?

4. Training, Documentation and Users

4.1. How many workstations do you have permanently available for training? Where are they located?

4.2. How many personnel perform training?

4.3. Do you have a section dedicated solely to training? Where and what does it consist of?

4.4. Have you in-house facilities to train users effectively in all the software you supply? If not, where is this carried out and by whom?

4.5. Do you run organized training courses for both specific packages and job functions? Give details here or separately of all the courses available and the costs involved.

4.6. Are you prepared to perform customized courses specific to the client's requirements? Give details, including costs.

4.7. What training is included with the system and maintenance contract?

4.8. What procedures ensure all documentation is updated and issued immediately any changes are implemented, such as revisions?

4.9. List all documentation supplied with a system (eg, manuals, bug reporting sheets, maintenance records, etc).

4.10. How are manuals indexed and cross-referenced (eg, by command, task, etc)?

4.11. Does the system support on-line documentation? Is this always kept up to date? How is it kept up to date?

4.12. Are users notified of all bugs within the system, given documentation for work-arounds regardless of if they have requested it, and informed when bugs have been fixed or patches are available? Give details.

4.13. Are there any subjects that are not currently covered within the documentation? Give details.

4.14. Is there an organized user group which has regular meetings? Do all users attend? What is its structure, size and the extent of its activities? How frequently do they meet? How is the operation of the user group(s) funded?

4.15. What market sectors are strongly represented within the user group?

4.16. What formal communication channels exist between the user group and the vendor?

5. Upgrading Quoted Configuration

5.1. What is the cost of increasing the number of workstations? List the additional hardware that is required to ensure consistent response. Show figures for increasing to four, five and six workstations.

5.2. What are the limitations for upgrading the quoted configuration (eg, maximum memory, mass storage, number of workstations, limits on peripherals)?

5.3. List all the additional software packages that are currently available and indicate their costs.

5.4. What software supported by your company does not operate on the quoted configuration? What hardware is needed to run these? What is the cost of this additional hardware?

5.5. What special provisions are made to assist users to upgrade their system to enable additional software to be used?

Appendix III:
Suppliers of Turnkey CAD Systems in the UK and the USA

The list that follows does not claim to be comprehensive or complete; given the delays inherent in the publishing process and the rapid rate of technological development in the CAD industry it would be futile even to attempt a full listing of all available systems. In fact the list is unashamedly subjective (and has a British bias), concentrating as it does on equipment which the staff of the Special Engineering Programme at Brunel University have either had occasion to use in the course of their own work or which they have found to be suitable for use in industry as a result of their studies and consultancy assignments.

It must also be emphasized that the list is a consolidated one, including some systems that have been found to be satisfactory only in a limited range of applications – which does not, of course, mean that they would necessarily be satisfactory in other applications. It should emphatically not be assumed that all the systems are comparable and can perform similar tasks – some certainly cannot.

The list should, therefore, be seen only as a starting point. It can be supplemented relatively easily by reference to the lists that are regularly compiled and published by a number of journals in the field.

In an attempt to provide some sort of key to the jungle of trade names which sometimes seems to be in danger of completely overgrowing the CAD industry, we have not only provided the names and current addresses of the companies but have also given, in each case, the name(s) of their principal software products as well as the type(s) of computer system which are currently being offered.

Applicon (UK) Ltd
Applicon Centre, Daw Bank, Stockport SK3 0EE
Telephone: 061 429 7227
Software: Bravo, EQUINOX
Computer: VAX
In the US: **Applicon Inc**
4251 Plymouth Road, PO Box 986, Ann Arbour, Michigan 48106
Telephone: 313 995 6000

Autodesk Ltd
South Bank Techno-Park, 90 London Road, London SE1 6LN
Telephone: 01-928 7868

Software: AutoCAD
Computer: Large range of micros including IBM PC and similar
In the US: **Auto Desk Inc**
2320 Marinship Way, Sausalito, CA 94965
Telephone: 415 331 0356

Auto-Trol Ltd
Neville House, 42-46 Hagley Road, Edgbaston B16 8PZ
Telephone: 021-455 7277
Software: 5000/7000 Series
Computer: APW (IBM PC); ARW(VAX); AGW(Apollo)
In the US: **Auto-Trol Technology Corporation**
12500 North Washington, Denver, Colorado 80133
Telephone: 303 452 4919

Calma Company
Beech House, 373-399 London Road, Camberley, Surrey GU15 3HR
Telephone: 0276 682821
Software: DDM
Computer: Apollo and VAX
In the US: **Calma Company**
527 Lakeside Drive, Sunnyvale, CA 94086
Telephone: 408 245 7522

Computervision
Central House, New Street, Basingstoke, Hants RG21 1DP
Telephone: 0256 58133
Software: CADDS4X, over 400 integrated applications packages for CAE/CAD/CAM
industries, Medusa
Computer: Own hardware, using also IBM PC, Sun & IBM 4300, VAX, Prime
In the US: **Computervision Corporation**
201 Burlington Road, Bedford, MA 01730

Daisy Systems (UK) Ltd
Berk House, Basing View, Basingstoke, Hants RG21 2HZ
Telephone: 0256 464061
Software: UNIX-based CAE for VLSI and System Design
Computer: IBM PC AT, Logician, VAX

Deltacam Systems Ltd
Aston Science Park, Aston Triangle, Birmingham B7 4AP
Telephone: 021-359 3659
Software: DUCT, 2-D Drafting, NC Programming & Management Systems, Dogs,
GNC
Computer: VAX, Prime, Apollo, Sun & Data General
American distributor: **Figgie Systems Management Group**
4420 Sherwin Road, Willoughby, Ohio 44094

Engineering Computer Services
Cooper House, Dam Street, Lichfield, Staffs WS13 6AA
Telephone: 0543 41751
Software: Graftek, Artech
Computer: HP9000, VAX & Gould
In the US: **Graftek Ltd**
717 Conestoga Street, Boulder, Colorado 80301
Telephone: 303 449 1138

Ferranti Infographics Ltd
Bell Square, Livingston, West Lothian EH54 9BY
Telephone: 0506 411583
Software: CAM-X
Computer: DEC VAX
American distributor: **Ferranti Infographics Div**
Ferranti Electro Optics Ind, 16812 Gothard Street, Huntington Beach, CA 92647
Telephone: 714 841 6812

Gerber Systems Technology Int Ltd
GST House, 4 Priory Road, High Wycombe HP13 6SE
Telephone: 0494 442121
Software: CAD/CAM system: SABRE-5000 series
Computer: HP 9000, MASSCOMP, IBM PC

Hytech Consultants
Chequers Parade, Wycombe Road, Prestwood HP16 0PN
Telephone: 9786 6071
Software: Micro/mini designer, electrical circuit design & printer circuit board design
Computer: VAX, Prime, IBM PC, IBM AT
American distributor: **Computer Dynamics Inc**
105 South Main Street, Greer, South Carolina
Telephone: 803 877 7471

ICL (UK) Ltd
Bridge House, Putney Bridge, London SW6 3JX
Telephone: 01-788 7272
Software: DIAD
Computer: PERQ
American distributor: **ICL Inc**
777 Long Ridge Road, Stanford, Connecticut 06902; PO Box 10276, Stanford,
Connecticut 06904
Telephone: 203 968 7200

Intergraph (GB) Ltd
Albion House, Oxford Street, Newbury, Berks RG13 1JG
Telephone: 0635 49044
Software: Mechanical, electronics, civil engineering
Computer: VAX and MicroVAX
In the US: **Intergraph Corporation**
One Madison Central Park, Huntsville, Alabama 35807
Telephone: 205 772 2000

International Research & Development Co Ltd
Fossway, Newcastle upon Tyne NE6 2YD
Telephone: 091-265 0451
Software: CADBIRD II
Computer: Available to run on various systems & supplied on PDP11, VAX, Data
General

Matra Datavision (UK) Ltd
Queen's House, Queen's Road, Coventry, West Midlands CV1 3HY
Telephone: 0203 21454
Software: Euclid
Computer: VAX
American distributor: **Matra Datavision Inc**
30 Commerce Way, Woburn, Massachusetts 01801

McDonnell Douglas Information Systems Ltd
Merion House, Guildford Road, Woking, Surrey GU22 7QH
Telephone: 04862 26761
Software: UNIGRAPHICS II
Computer: DG, VAX & IBM
In the US: **McDonnell Douglas Manufacturing Industry Systems Company**
PO Box 516, St Louis, Missouri 63166
Telephone: 800 325 1551

Norrie Hill Ltd
Apex House, London Road, Bracknell, Berks
Telephone: 0344 481717
Software: Source 36
Computer: HP9000

British Olivetti Ltd
PO Box 89, Olivetti House, 86-88 Upper Richmond Road, London W15 2UR
Telephone: 01-785 6666
Software: GTD (CAD), GTL (CAM), GTX OMS for manufacturing control software
Computer: Olivetti
American distributor: **Docutel/Olivetti Corp**
5615 Highpoint Drive, Irving, Texas 75062
Telephone: 214 258 5400

Pafec Ltd
Strelley Hall, Strelley, Nottingham NG8 6PE
Telephone: 0602 390649
Software: Dogs & Boxer
Computer: Apollo software for range of systems
American distributor: **Pafec Ltd**
6855 Jimmy Carter Boulevard, Suite L/1200, Norcross, Georgia 30071

Prime Computer (UK) Ltd
Primos House, 2 Lampton Road, Hounslow, Middlesex TW3 1JW
Telephone: 01-572 7400
Software: Medusa, SAMMIE, Product Design Graphic System (PD 6S)
Computer: Prime 50 Series
In the US: **Prime Computer Inc**
Prime Park, Natick, Massachusetts 01760
Telephone: 617 655 8000

Racal-Redac Ltd
Green Lane, New Town, Tewkesbury, Gloucestershire GL20 8HE
Telephone: 0684 294161
Software: CIEE
Computer: Apollo, IBM PC, Micro VAX
American distributor: **Racal-Redac Inc**
Liberty Way, Westford, Massachusetts
Telephone: 617 692 4900

Robocom Ltd
Clifton House, Clifton Terrace, London N4 3TB
Telephone: 01-263 8585/3388 or 01-272 8417 (4 lines)
Software: Robosystem; Bitstik; RoboCAD-PC
Computer: Apple 11 + /e; BBC Model B/B + ; IBM PC/AT/XT and various
compatibles

American distributor: **Chessell Robocom Corporation**
111 Pheasant Run, Newtown, Pennsylvania, PA 18940

Tangram Computer Aided Engineering Ltd
5 Siddeley Way, Royal Oak Industrial Estate, Daventry, Northamptonshire NN11 5PA
Telephone: 0327 705026
Software: SWIFT-CAD, SWIFT-NC, SWIFT-CAM, SWIFT-FE
Computer: Apollo, Digital, Data General, IBM-AT, Prime, Systime

Select Bibliography

CAD/CAM

Aleksander, I., ed. *Computing Techniques for Robots* Kogan Page, London, 1985 – ISBN 0 85038 934 8

Besant, C. B. *Computer-aided Design and Manufacture* Ellis Horwood, Chichester, 1980 – ISBN 0 85312 117 6

Groover, M.P.; Zimmerman, E.W. *CAD/CAM: Computer-aided Design and Manufacturing* Prentice-Hall Inc., New Jersey, 1984 – ISBN 0 13 110130 7

Milner, D. A.; Vasiliou, V. C. *Computer-Aided Engineering for Manufacture* Kogan Page, London, 1986 – ISBN 1 85091 093 6

Design

Hubka, V. *Principles of Engineering Design* Butterworth, Guildford, 1982 – ISBN 0 408 01105 X

Medland, A.J. *The Computer-Based Design Process* Kogan Page, London, 1985 – ISBN 0 85038 937 2

Electronics

Wolfendale, E. *Computer-Aided Design of Electronic Circuits* Illife Books, London, 1968

Graphics

Gardan, Y.; Lucas, M. *Interactive Graphics in CAD* Kogan Page, London, 1985 – ISBN 0 85038 798 1 (revised edition)

Foley, J. D.; Van Dam, A. *Fundamentals of Interactive Computer Graphics* Addison-Wesley, London, 1982 – ISBN 0 201 14468 9

Pitts, G. *Techniques in Engineering Design* Butterworth, Guildford, 1973 – ISBN 0 408 70438 1

Mathematics

Encarnacao, J. L.; Schlechtendazhl, E.G. *Computer-Aided Design* Springer-Verlag, New York – ISBN 3 540 11526 9

Faux, I. D.; Pratt, M. J. *Computational Geometry for Design and Manufacture* Ellis Horwood, Chichester, 1979 – ISBN 0 85312 114 1

Giloi, W. K. *Interactive Computer Graphics* Prentice-Hall Inc., New Jersey, 1978 – ISBN 0 13 469189 X

Mullineux, G. *CAD: Computational Concepts and Methods* Kogan Page, London, 1986 – ISBN 1 85091 110 X

Newman, W. M.; Sproull, R. F. *Principles of Interactive Graphics* McGraw-Hill International, New York, 1982 – ISBN 0 07 046338 7

Social Effects

Encarnacao, J. L.; Torres, O. F. F.; Worman, E. A. *CAD/CAM as a Basis for the Development of Technology in Developing Nations* North Holland Publishing Company, Oxford, 1981 – ISBN 0 444 86320 6

Warner, M., ed. *Microprocessors, Manpower and Society* Gower Publications, London, 1984 – ISBN 0 566 00631 6

Index

Page numbers in italics refer to illustrations.